Praise for *Tides*

"Jonathan White's tidal explorations drew me in with just the right mix of science, history, and storytelling, propelled throughout by the author's infectious curiosity and sense of wonder. Beautifully written, impeccably researched, and filled with unexpected connections and discoveries, *Tides* is a splendid book—highly recommended."
  —Thor Hanson, author of *Feathers* and *The Triumph of Seeds*

"I loved this book. As a physical oceanographer, I understand tides from a scientific viewpoint. I enjoyed learning the history of tidal theories, the spiritual meaning the tides have for many people around the world, and how the tides affect so many places in such varied ways. I recommend it to both scientists and nonscientists alike."
  —Sally Warner, Ph.D. in physical oceanography,
    Oregon State University

"*Tides* is an enriching meditation on the motions, eccentricities, and ebbs and floods of the 71 percent silent majority covering the planet's surface. In clear, poetic verse, White paints the ocean for what it really is: less a static mass and more a living, breathing being swayed by the rhythms of the universe that, in ways large and small, interconnects all life."
  —James Nestor, author of *Deep: Freediving, Renegade Science, and What the Ocean Tells Us about Ourselves*

"Let me be clear. This is one of the most fascinating, engaging, relevant, and impeccably brilliant books I have ever read. It has profoundly changed my sense of the earth, the oceans, the sky, and how they are deeply interwoven with the course of human thought and history."
  —Richard Nelson, author of *The Island Within*

TIDES

The Science and Spirit of the Ocean

≈

JONATHAN WHITE

Foreword by
Peter Matthiessen

TRINITY UNIVERSITY PRESS
SAN ANTONIO

Published by Trinity University Press
San Antonio, Texas 78212

Jacket design by Ann Weinstock
Book design by BookMatters, Berkeley
Cover art: moon, © Colourbox.com; water,
   exsodus/123rf.com

ISBN-13 978-1-59534-805-0 hardcover
ISBN-13 978-1-59534-806-7 ebook

Trinity University Press strives to produce its books
using methods and materials in an environmentally
sensitive manner. We favor working with manufac-
turers that practice sustainable management of all
natural resources, produce paper using recycled stock,
and manage forests with the best possible practices
for people, biodiversity, and sustainability. The press
is a member of the Green Press Initiative, a nonprofit
program dedicated to supporting publishers in their
efforts to reduce their impacts on endangered forests,
climate change, and forest-dependent communities.

The paper used in this publication meets the minimum
requirements of the American National Standard
for Information Sciences—Permanence of Paper for
Printed Library Materials, ANSI 39.48–1992.

CIP data on file at the Library of Congress

21   20   19   18   17 | 5   4   3

For Donna and Matthew,
my loves

The sea has many voices,
Many gods and many voices.

– T. S. Eliott

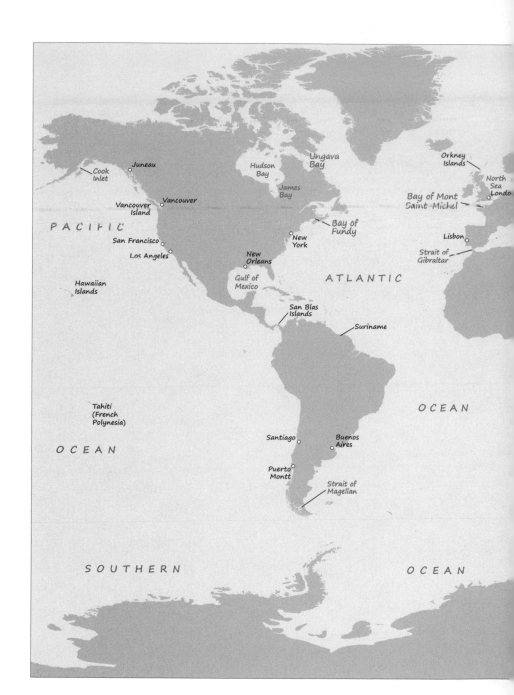

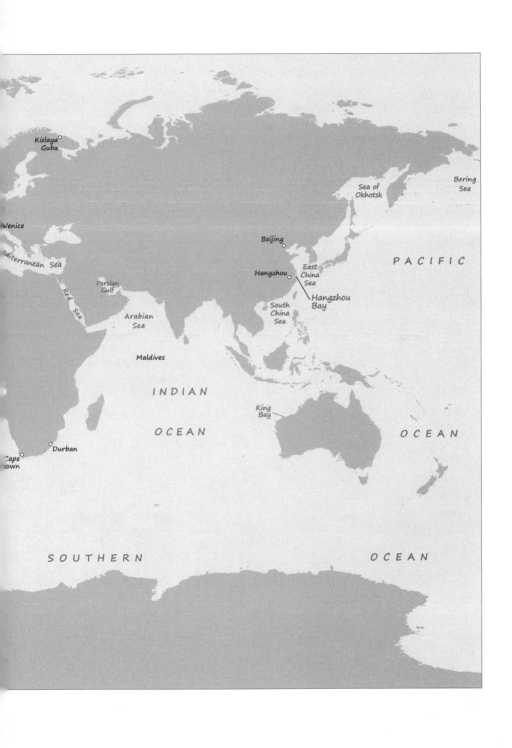

Contents

Foreword

The mysterious workings of the tides are among my own earliest memories. On childhood days on a sand edge of Fishers Island, off the New England coast, I would sit mystified by the water's unaccountable retreat just as it touched the entrance channel into my sand castle. In boyhood summers on fishing excursions with my father off Race Point at the west end of that island, I observed with dread the relentless shifting of dark muscled waters all the way west to the lighthouse at Little Gull Island on Long Island's North Fork, three miles away. Through this great tideway called the Race, twice every day, the broad sound between Long Island and Connecticut empties itself into the Atlantic, only to return on the flood tide six hours later. The sheer might of this current, with the groan of fog horns and the distant gong of the gaunt bell buoy at Valiant Rock, had such a grip on my imagination that I named my first novel after the Race Rocks Lighthouse for no better reason than its evocation of the fathomless power of existence.

> The Race was forever in a state of change, and its faces were gray and blue and black, and red with torn menhaden when the bluefish ran, and scarred with white. Its force gave him the feeling that nothing lost there was recoverable but would fade into some deep oblivion of anonymous salt tatters and marine rust. Once he himself had clutched at a green

glass net float from another ocean only to see this unsinkable thing drift down from the light into the shadows of water rushing hard upon Race Rock . . .

South and west across the Race, passing south of Little Gull, then westward down the long rough reach north of Gardiner's Island toward the South Fork and the old whaling port of Sag Harbor: only rarely in six decades as a boatman have I made that voyage west or east without first consulting with the tide and current tables of my coast pilot.

On the South Fork, as a young writer, I worked the seasons for three years as an ocean haul seiner and scalloper and, in summer, as a charter boat captain out of Montauk Harbor—and those work days were all at the beck and call of tides. For the past fifty years I have lived year-round across the South Fork from Sag Harbor in what was once a small farming community behind the dunes where, on certain winds, the explosion of breaker is clearly audible; sometimes, in a storm, the rush of ocean, obliterating the rumble of the surf, sounds like an oncoming avalanche, threatening to overwhelm the world.

~~

How fitting, then, that on a stormy evening I find myself immersed in *Tides,* an excellent new book on the earth's tides by Jonathan White, a veteran mariner who for many years at the helm of his schooner, *Crusader*, has navigated the strong currents of the Pacific Northwest archipelago, notably that monumental passage that extends from the state of Washington and Vancouver Island to the glacial coast of southeastern Alaska.

Happily, he is also a fine writer and reporter, pursuing his elusive subject all the way from southern China to Ungava Bay on the Arctic coast of North Quebec, where a differential of over fifty feet is among the world's most extreme tides; here, near the village of Kangiqsujuaq, he follows an Inuit shellfish forager down a hole chopped through the thick onshore ice that forms a roof over the exposed mussel beds in the eerie chamber left behind by the ebbed tide.

White's book begins sensibly enough with fascinating discussions of the nature of tides and the relation of their moon-led cycles to the cycles of all forms of life, including man. The precise timing of certain bird migrations linked to the breeding cycles and abundance of food species is well known—the dependence, for example, of the northbound red knot and other sandpipers on the myriad eggs of the horseshoe crab at Cape May (New Jersey) in the spring and the explosion of small mudshrimp on the tide flats of the Bay of Fundy in fall migration. Years ago I was lucky enough to see a green turtle haul herself ashore in the Red Sea and lay her eggs in the warm sand above the tide line, the hatch perfectly timed to that high tide months later in the dark of the new moon, when the new turtles will dig their way free of the wet sand and rush toward the safety of the sea.

White also explores the scientific history of tide theory, its lore, and the strange nature of waves. He investigates tidal currents—up to a scary eighteen knots in the archipelago of the Pacific Northwest Coast—and describes the tidal bores that in certain rivers all around the world may surge upriver on the incoming tide as a single dangerous wave. (I think of the Eilanden River bore on the south coast of New Guinea, which on its return downriver overturned the motor raft of young Michael Rockefeller on our 1961 expedition to the war tribes of the highlands and led to his controversial disappearance.) White even journeys to China's Qiantang River to observe the largest of all bores—which attains heights of more than twenty feet and necessitated a tide table predicting its arrival times as early as the mid-eleventh century.

Finally, he studies the intricacies of so-called resonance—now recognized as a primary factor in the largest tides as well as the smallest—and the hundreds of cycles of sun and moon that influence any given tide. He visits the San Blas Islands near Panama and their Kuna Indians, who will lose their homeland of 350 low-lying islands if sea levels rise three feet, as conservatively predicted, in the next fifty years. Lastly and importantly, in a time of climate change and rising seas, he explores the exciting potential of the tides as a source of clean energy. (Himself a surf swimmer, he knows the strange "heaviness" of a breaking wave and its wash along the ocean beach after a storm.) Some years ago, I read somewhere that harnessing a single wave here on the hundred-mile ocean beach on the South Shore would provide electric power for this whole densely developed island for twenty-four hours.

Jonathan White provides us in this fine, fascinating book with a clear understanding of the infinitely complex and wild nature of our planet's tidal forces in all their mystery and beauty. He is to be warmly congratulated.

Peter Matthiessen
Sagaponack, New York, 2014

In Deep

An Introduction

～ My interest in tides springs from a fascination with the ocean. I grew up on the southern California coast, surfing, diving, sailing, fishing. I built a twenty-six-foot sloop after college and sailed it for a couple of years in the Atlantic and Caribbean, making several offshore passages. In the early 1980s, at twenty-five, I bought a leaky old sixty-five-foot wooden schooner, *Crusader*, and founded a nonprofit educational organization, Resource Institute. For eleven years we sailed *Crusader* off the Northwest Coast, from Seattle to Alaska, around Vancouver Island and Haida Gwaii. We conducted weeklong seminars afloat on topics ranging from natural history, photography, and whale research to psychology, music, poetry, and Northwest Coast Native art, culture, and mythology. Among the seminar presenters were Peter Matthiessen, Lynn Margulis, Gary Snyder, Paul Winter, Robert Bly, Art Wolfe, Gretel Ehrlich, and Roger Payne. Six or eight participants from across the country—sometimes from around the world—would join us at a coastal town, and we'd sail off, often not seeing another human settlement until the seminar's end.

It was a wonderfully adventurous eleven years, but one of the not-so-wonderful adventures was going aground on a large tide in Alaska's Kalinin Bay.

I'd been aground before. In fact, I've been aground more often than I care to admit, because I misjudged the tide or misread a tide table. Over the years, I frequently bounced on and off rocks and sandbars, sometimes getting stuck and having to wait for a rising tide to lift me off. Most groundings were innocuous—sometimes agonizing and almost always embarrassing, but not damaging to the boat. I always looked at these experiences as a reminder of who's really in charge. Clearly, not me.

Where I live, on the Northwest Coast of North America, sailors need to watch the tides as actively as they trim sails. Although we think of ourselves as sailing across the ocean's surface, we are also sailing down the low tide's valleys and up the high tide's mountains. Hanging on as a tidal passenger isn't always as easy as it might seem. Kalinin Bay taught me that.

It was August 1990, and twelve of us, including three crew members, had been sailing for a week on the isolated coasts of Kruzof and Chichagof islands with anthropologist and writer Richard Nelson for a seminar called "Nature, Culture, and World View." We had anchored for the night in the tiny bight of Kalinin Bay on the north tip of Kruzof Island. The next day we planned to hike the island before heading into Sitka to finish the trip.

The night was star-filled and quiet when we went to bed, but in the early morning hours I was awakened by a howling gale. I pulled on my rubber boots and climbed on deck to make sure everything was okay.

It wasn't. We had dragged our anchor across the bay and were aground in the mud. With a lump in my throat, I grabbed the tide chart from the pilothouse. If the tide was low and about to turn, we would be in luck. *Crusader* would refloat easily with the flood, just as she had done before. If the tide was high and on its way down, however, it'd be disastrous. With a fourteen-foot tide, the water would literally disappear from under *Crusader*, leaving us stuck in the mud.

Flipping to the pages for Sunday, August 19, I read "high tide 05:00, low tide 09:50." I read it several times. Really? If this was right, we had gone aground at peak high tide—the worst possible time. There wouldn't be enough water in Kalinin Bay to refloat *Crusader* for another nine or ten hours, and the whole time we'd be settling more deeply into the mud. In the worst case, we'd be sunk too deeply to get out at all.

The crew and I woke everyone and shuttled them in small boatloads to a tarped shelter onshore. Richard stood by with a rifle in case of bears. The wind was gusting at more than forty knots, pouring off the mountains and whipping the bay into frothy streams. It ripped one of our kayaks from *Crusader*'s deck and flipped it like a toothpick across the bay. Driving rain stung my eyes. The nearest help, in Sitka, was at least four hours away by boat. As a general rule, a captain doesn't call for assistance unless it's direly needed. But I wouldn't know if we needed help until the tide was rising, and by then it might be too late for any help to make a difference.

Over the hours, I watched *Crusader* drop like a fatally wounded animal, first to her knees, then all fours, and finally onto her side. She filled chest-high with water. When the tide reversed, all seventy tons of her were stuck in the mud and didn't want to come back up.

Crew member Lela Hilton and I stayed on the boat, trying to rescue valuables, swimming in the flooded cabins. I worried about the engine, a tractor-sized Detroit diesel. Maneuvering in tight anchorages or marinas would be impossible without it. I watched the rising water lick at the oil pan, then swallow it. The batteries were next, then the fuel injectors. By midmorning the engine's whole iron mass had vanished under the turbid water.

Food, shoes, and fruit seemed to delight in their newfound buoyancy. Packages of raisins and mint cookies bobbed here and there. Herring balls of rice swirled above the quarter berths. And the books. The books! Two hundred or more, swelled with ocean, took up dancing partners with the apples and pears and tennis shoes. I'll never forget the titles that floated by: Ed Rickett's *Between Pacific Tides*, Robert Bly's *When Sleepers Awake,* Ram Dass's *How Can I Help?* Lela managed a smile as she held up a soaked copy of Maxine Kumin's *In Deep.*

The sixty-five-foot schooner *Crusader* aground at low tide in Kalinin Bay, Alaska.

Finally I realized the boat was at risk of becoming a total loss, and I called for help. The Coast Guard sent a helicopter with three large pumps. The pilot radioed that they'd like to stay and help, but two other boats were in trouble up the coast. They left gas for the pumps, which we started immediately, placing one in the forward cabin and two in the engine room.

Even with the pumps gushing 350 gallons a minute, the effort seemed futile. There was no sign of gain. Sometime in the late morning, I paused on the top rung of the engine room ladder, feeling the terrific cold for the first time. I was exhausted. We all were. The group on land was rescued midmorning by a local fisherman who had heard my radio call for help. I was glad to know they were safely on their way back to Sitka. Lela and I stayed behind, just the two of us, having gotten all the help we could expect.

It was time to let *Crusader* go. I wondered what the insurance company would need. What should I take with me as I left her for the last time?

Then something shifted. Lela and I watched, transfixed, as the cap rail and bulwarks reappeared. Long blades of eelgrass disentangled from the deck and slipped back into Kalinin Bay. Released from the muddy bottom, *Crusader* was floating upright in ninety seconds.

Lela kept the pumps working while I drained seven gallons of seawater from the engine's main oil chamber and three more from the gearbox. Lela helped me pull the valve covers and injectors, exposing the engine's soaked interior. After we jerry-rigged the fuel and electrical systems, we were able to crank the engine. The first revolutions spat out a few more quarts of seawater, but with each revolution there was less and less water, until the engine room was finally filled with a fine mist of pure diesel. At 2:00 a.m., almost twenty-four hours after we had gone aground, I eased the six injectors back in their sleeves, tightened them in, said a prayer, and pushed the starter button.

I stood on the aft deck minutes later, watching the light blue exhaust float over the bay. The wind had died and the bay was once again cradling that giant Alaskan silence. We slept for a couple of hours and then motored into Sitka the next morning. With a crew working around the clock to clean and repair the cabins, three days later *Crusader* sailed off on the next seminar.

⇌

After Kalinin Bay, I vowed to learn more about the tides. I couldn't have prevented the grounding simply by knowing how the tide worked, but nevertheless I figured it was time to learn more.

I knew the moon had something to do with it, but what? I thought I'd find my answer in a book or two, but the more I read, the more complex and mysterious and poetic the subject became. I learned, for example, that planetary motion, which governs the tides, is not at all simple or regular. It's full of eccentricities. The sun, moon, and earth don't orbit in perfect circles. At times they're closer to one another and at times farther away. They speed up and slow down. They wobble and yaw and dip and veer, and each time they do, it translates into a tidal event on earth.

There are hundreds of these eccentricities, each calling out to the oceans—some loudly, some faintly, some repeating every four hours and others every twenty thousand years.

How the oceans hear these heavenly calls is another story. Some oceans hear only a single voice; others hear a chorus. The calls are heard differently in Boston than San Francisco, London, or Shanghai. They're even heard differently in two bays several miles apart. The Atlantic is strongly tuned to the moon; the Pacific is tuned more to the sun.

The more I found out, the more I wanted to know. I wrote a piece on Northwest Coast tides for *Orion* magazine. I went to China under contract with *Natural History* magazine to write about the Qiantang River's tidal bore, a wave that roars upriver on every flood tide—sometimes reaching twenty-five feet in height. I came across so many fascinating stories about the tides during these projects that five years ago I embarked on this book.

Had I known how wet and confused I would become on this journey, perhaps I would have hesitated at its edge, or at least donned a life jacket before wading in. When I wallowed—and I often did—I found consolation in knowing I was in the best of company. Aristotle was befuddled too. He lived 2,300 years ago on the Mediterranean, where the tidal range is only a few inches. Yet in the Euripus Strait near his home, he was perplexed by a strong reversing current. Aristotle suspected it had something to do with the tide, but he couldn't figure out how or why. Three different historical accounts suggest that Aristotle's frustration reached such intensity that he ended his life by throwing himself into the sea, exclaiming, "Comprehend me, since I cannot comprehend thee!"

Two thousand years later, Newton ignited a revolution when he identified the tide's cause as that ghostly force called *gravity*. In a world that was desperately trying to shake free of ancient mystical views, accepting the existence of an unseen force was like believing in witchcraft. How could a force push things around—over such great distances, no less—without showing itself? Gravity was spooky, even to Newton. He called it "a most disgraceful thing" but plunked it nonetheless into the center of his revolutionary tide theory.

Modern-day tide theory has come a long way since Newton but hasn't rid itself of mystery. In my research, I met dozens of accomplished oceanographers who flatly admitted that the tides were too complex for any one person to fully understand. When I asked renowned tide modeler David Greenberg to describe an "aha" moment, he sat uncharacteristically quiet for a while. We were in his office at the Bedford Oceanographic Institute in Nova Scotia, not far from the Bay of Fundy, where the tides have been recorded at fifty-four feet, six inches—the largest, along with the Arctic's Ungava Bay, in the world.

"I don't have 'aha' moments in this field," he finally answered, "only 'oh god' moments when I find something that makes no sense."

≈

Before Newton introduced gravity and a unified picture of planetary motion, the ancients put forth tide theories based on myth, astrology, practical observations, and religious beliefs. They used the conceptual tools they had, as we do today, to make sense of their world.

The Maori of New Zealand believed the tides rose and fell at the whim of a woman-god who lived on the moon. The Chinese envisioned the Milky Way as a great waterwheel, filling and emptying the oceans as it churned. In many cultures, there was a perceived "secret harmony" between the tide and human life. Flooding tides were a time of exuberance, prosperity, conception, birth. It was a time to make butter and sow clover. Ebbing tides were a time of melancholy, harvest, death. A woman's menstrual cycle was the tide's ebb and flow within her body.

The tide's long, steady inhale and exhale is suggestive of a living being. Some thought the tides were the breathing of Gaia, the earth itself, and others suspected it was a large beast. Leonardo da Vinci was convinced of the latter and tried to calculate the size of its lung.

Whatever speculation about the tide's cause or secret harmony, the earliest coastal people most certainly accumulated vast practical knowledge about daily, weekly, monthly, and yearly tide patterns. They needed this information to survive, to know where and when to pick mussels or gather seaweed in the intertidal zone, when to launch a boat and how to take advantage of currents. You only need to paddle against the tide once to be convinced never to try it again. From a practical point of view, it didn't matter if the tide was caused by a waterwheel or a beast or a god; it only mattered that daily survival was easier if you observed the tide and worked with it, not against it. The more you knew, the better.

This was true for all coastal peoples but especially for the canoe cultures that settled in extreme environments like Tierra del Fuego, the Arctic, or the northern coasts of North America. These places, with their large tides, ferocious currents, and frigid water, were bountiful but mercilessly demanding. The wrong move in a canoe loaded with fish or clams could be fatal. The knowledge these people acquired was passed in stories and hands-on teachings from generation to generation, for thousands of years. Contemporary written records have captured precious little of this traditional knowledge.

I learned a little about the depth of indigenous knowledge when I went hunting several years ago with Lukasi Nappaaluk, an Inuit elder from the Nunavik region of northern Quebec, Canada. We had hiked across the tundra toward Hudson Strait, where we planned to launch a small boat. As we approached the shore, I guessed that the tide was at midrange.

"It's flooding," Lukasi said.

It was a quiet beach, and to my eye there was no indication whether the tide was flooding or ebbing.

"How do you know?" I asked.

"Fuzz," he answered.

He would have left it at that, but I persisted.

"What fuzz?"

"When the tide is flooding, it picks up dust and pollen and insect larvae from the beach that sits on top of the water like a blanket. It doesn't happen when the tide is dropping."

For Lukasi, the fuzz tells him which way the currents are moving, how fast, and for how long. It tells him when to launch his boat and what course to steer. It tells him whether he should round Point Frontenac looking for seals or head for Cape Neptune to fish for char.

Today, I notice fuzz around Washington State's San Juan Islands, where I live. I see it from my car as I drive. In the late afternoon sun, it looks like velvet.

⁓

The tides rise and fall constantly on all the world's coastlines—more than 370,000 miles of them. In places the range is small, such as the Mediterranean, the Gulf of Mexico, and the South Pacific Islands; in other places it's large, as in northwestern Australia, Patagonia, the United Kingdom, and northeastern Canada. Large or small, the tide is always on the move, swelling against one coast while shrinking from another. It never begins and it never ends.

As I write this, I imagine what the tide might be doing around the world right now. Where I live, it's low, and oyster grower Nick Jones is out on the flats with his crew, repairing beds, planting spat, and harvesting oysters for market. While he hauls burlap bags up the intertidal zone, a large heron in Suriname lifts its blue-black body from the roost and squawks loudly as it wings toward the coast. Its destination is miles from sight, but an internal clock signals that the tide is low, and a meal of mussels, urchins, and crabs awaits. As the bird alights on a seaweed-covered rock, David Plunkett dons a white apron and opens the sluice gates at the Eling Tide Mill, south of London. In minutes, the sixteenth-century mill, the oldest operating in the world, moans to life, and warm yellow flour spills from the grinding wheels. Later, when the tide is low and the wheels are again still, the whole-meal flour will be delivered to bakeries in Totton and Bitterne to be made into cookies and bread.

When those cookies are in the oven, Lukasi Nappaaluk might be cutting a hole in the Arctic ice and preparing to drop below into the womblike hollows left behind by an extreme low tide. In that ice-roofed dark cavern, he harvests fresh blue mussels for dinner.

Thousands of miles away, in Venice, Italy, a young couple sits at a small table, their hands entwined. In the fading evening light, they gaze at each other and the full moon rising over Piazza San Marco. There's nothing unusual about this romantic scene, except that the couple's chairs are sitting in sixteen inches of seawater. It's *acqua alta* in Venice. High tide has jumped the seawall and flooded San Marco Square, something that has happened occasionally for centuries but more frequently in recent years—about a hundred times in 2015—as global sea levels rise. Though these floodings pose a serious threat, Venetians are accustomed to it. Visitors walk on raised wood planks and shop in rubber boots. The young couple can still order a cappuccino, but the server who brings it out is wearing hip waders.

~~

Acqua alta might not contribute to Venice's romance, but a full moon certainly does. Who can gaze upon this bright companion and not feel inspired? No wonder our ancestors thought of it as mother, father, god, death, renewal. They understood, long before science revealed the details, that the moon and tides are connected, mirroring each other as they rise and fall, grow and shrink, die and are reborn.

Cultures of the not-too-distant past played music and danced to the moon, celebrating fertility, ripeness, love. *That* moon is the same one we see today, the one that reminds us of where we came from. A century ago, Yeats wrote

What is there in thee, Moon!

that thou shouldst move
my heart so potently?

The moon may move our hearts today, but her first love was the ocean, stirred billions of years ago. The attraction may have grown stronger or weaker through time, but the affair has never ended. Like any relationship, it has complexities and baggage. The moon calls out to the earth's oceans in the form of gravity, and the oceans call back, their pulsing energy holding the moon close while also pushing it away. It's a dance performed celestially, with partners on a floor hundreds of thousands of miles across. It brings new meaning to the concept of a long-distance relationship. And, from the perspective of a human life, the dance never had a beginning or an end.

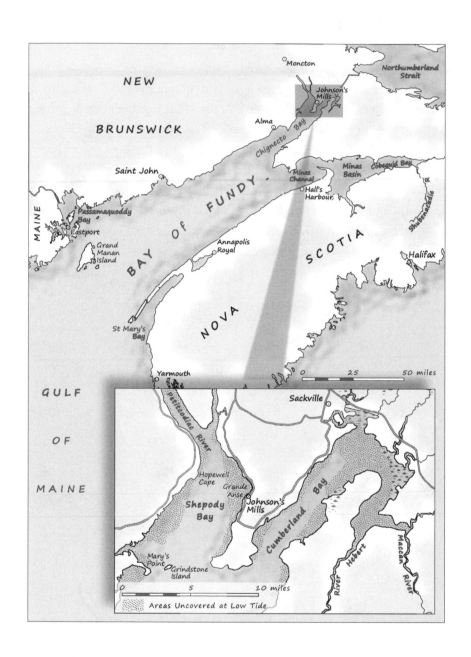

NEW

BRUNSWICK

Moncton

Northumberland
Strait

Johnson's
Mills

Alma

Chignecto Bay

Saint John

Minas
Channel

Minas
Basin

Cobequid Bay

Hall's
Harbour

MAINE

Passamaquoddy
Bay

Eastport

BAY OF FUNDY

SCOTIA

Shubenacadie

Grand
Manan
Island

Annapolis
Royal

Halifax

NOVA

St Mary's
Bay

25   50 miles

GULF

Yarmouth

Petitcodiac River

Sackville

OF

Hopewell
Cape

Grande
Anse

MAINE

Shepody
Bay

Johnson's
Mills

Cumberland Bay

Mary's
Point

Grindstone
Island

River Hebert

Maccan River

0   5   10 miles

Areas Uncovered at Low Tide

1 The Perfect Dance

Birds and Big Tides in the Bay of Fundy

> It's a strange thing that most of the feeling we call religious . . .
> is really the understanding and the attempt to say that man
> is related to the whole thing, related inextricably to all reality,
> known and unknowable. . . . It is advisable to look from the
> tide pool to the stars and then back to the tide pool again.
>
> — *John Steinbeck, The Log from the Sea of Cortez*

When a late summer low tide slips across the Bay of Fundy, it reveals miles of glistening mudflats. All that mud is filled with *Corophium*, a mudshrimp about the size of a rice grain. And every August, just as the mudshrimp population peaks, 700,000 semipalmated sandpipers descend onto these tidal flats. This bird is so small it fits in the cavity between two clasped hands, and so light you wouldn't notice it was there.

Having flown 900 miles nonstop from the Arctic to this great Canadian bay, the sandpipers land weak and skinny. But not for long. In eight days on the tideline, they double their weight with mudshrimp. They'd better. If they don't consume enough calories for the nonstop 2,500-mile trek over the Atlantic to their South American wintering grounds, they'll drop out of the sky somewhere midocean. "It's like taking off in your car across a wilderness, knowing there is only one gas station—exactly halfway. If you miss it, or if you don't fill your tank to the brim before leaving, you simply won't make it," Peter Hicklin tells me.

Hicklin is a retired Canadian Wildlife Service biologist. He is recounting his thirty years of shorebird research—including two life-threatening accidents—as I steer a rental car between potholes and soft red clay. It's early August, and we're headed to Johnson's Mills, a large tidal flat in Shepody Bay, New Brunswick. The roadside is overgrown with alder and black spruce, but at times it opens to sweeping green fields of hay and corn, punctuated by large gambrel-roofed barns and crisp white Victorian farmhouses. The muddy-brown sea, Shepody Bay to the west and Cumberland Basin to the east, is never far from sight.

"In the early 1970s we knew almost nothing about this place," says Hicklin, his accent hinting at his French Acadian heritage. "The tidal flats were considered a dangerous wasteland—no one went out there. We knew Fundy had the largest tides in the world, but we didn't know until recently that they have shaped everything that lives here, from the careers of scientists and fishermen to birds and fish and mudshrimp. If it weren't for the tides, none of this would be here. Hell, I wouldn't be here."

≈

"Shhhh," whispers Hicklin as we park and walk to the beach. The tide is high, almost at its peak; small wavelets from a pleasantly cool onshore breeze lap the gray rocks and sand of the upper shore. People in light parkas and shorts hover by the roadside shrubs; they carry cameras with long lenses, spotting scopes on tripods, and sophisticated video equipment. Children lie on their stomachs in fireweed and mustard, propping up binoculars. They whisper and silently swat mosquitos, repositioning themselves for the optimum view.

Like me, they have timed their trips from New York and Ontario and Wisconsin to arrive at just this hour. We—the birds and the people—are on two different migrations: the birds on wings across Arctic tundra, we in automobiles across paved highways, together balancing for a moment on the crest of a Fundy high tide.

Semipalmated sandpipers above the Bay of Fundy.

If I had somehow floated into the bay on driftwood from points south, I might have drifted in and out again without noticing a thing. But Hicklin's "Shhhh" and the dedicated birders direct my focus. At first I don't even see bird shapes, and then I realize the entire beach is shimmering with life. Other than a few small patches of sand and rock, the whole beach is a vibrating carpet of tiny, beautiful sandpipers. Thousands of them.

Occasionally the flock takes to the air. As one, they lift with the soothing hiss of a wave retreating over loose pebbles. Aloft, they swirl and dive and bank. One moment their white underbellies are turned toward the shore, and the flock disappears against the sunlit sky; as they turn again, a wave of gray rushes through the group and they appear as smoke—wispy and shifting, twisting and hovering. They are perfectly choreographed dancers—graceful, supple, arching, rhythmic. Watching them, we are speechless. No music plays, but the dancing flock is as musical as anything I have seen.

～

I first journeyed to the legendary Bay of Fundy in 2011. With a tidal range (the vertical measurement between consecutive high and low waters) of 54.6 feet, it competes with Ungava Bay, 1,200 miles north, for the world's largest tide. I drove the 350-mile coast of Nova Scotia and New Brunswick, the Canadian provinces that border the bay, meandering from the lobstering and scallop-fishing regions at the bay's mouth to the upper inlets of Chignecto Bay and Minas Basin, where the transition between tide and land blurs.

In the fishing community of Alma, I saw the incoming tide lift the fleet with a gentle hand. Because boats in these harbors are afloat only a few hours either side of high tide, I mused at how fishermen must schedule their comings and goings. Terry Rossiter, a fisherman working out of Alma for thirty years, told me from the tailgate of a red pickup, "Most people live by the sun, but if you fish Fundy you live by the moon and tides. You're on a completely different schedule than everybody else." The tide, mostly driven by the moon, changes by about an hour every day. So if Rossiter leaves the dock tomorrow at three in the morning, the next day it will be four, and a week later it'll be almost noon. "We eat and sleep on that schedule," he says. "When the tide's flooding, whatever hour it is, the crews go to their boats, load things on—food and bait—get things ready. We have a failsafe mark on the wharf, so when the washboard hits that mark, we go, ready or not. If you miss the tide, you miss the whole day."

Rossiter is one of hundreds who live by the moon and tides of Fundy, one of millions who live this way worldwide. In upper Fundy, weir fishermen take advantage of the tides by stretching a net between two poles at low tide. Cod and shad are swept in by the flooding tide and picked during the next low. At Dark Harbour on Grand Manan Island, harvesters collect the edible seaweed dulse during the lowest tides of spring and summer. These occur roughly every two weeks, coinciding with the new and full moons. Pickers in Dark Harbour call these *dulsing tides*, gathering night or day to handpick the ruddy, fingerlike fronds. They flourish in the lowest part of the intertidal zone, which is the area between high and low water. Hauled to shore in burlap bags, the wet harvest is spread to dry in the sun. Dark Harbour sits at the bay's mouth and is the dulse capital of Fundy, exporting 75,000 pounds a year of this crispy, salty, high-mineral snack. Some goes to the United States, Europe, and Australia, but much of it stays in Canada, to be eaten raw or pan-fried as chips, baked in the oven and topped with cheese, or sprinkled on salads.

A growing population of Fundy scientists study the tide's far-reaching and mysterious workings. Biologists examine how marine animals synchronize their life cycles with it. Geologists look at how currents shape the ocean bottom. Sedimentologists monitor suspended clays and sand that are stirred and redeposited. Oceanographers research tide anomalies—which is almost an oxymoron, since anomalies abound. Engineers improve devices that generate electricity from the tides.

Fundy's tides shape the bay's landscape and culture. What happens here happens to some extent everywhere. More than half the world's population lives on or near the coast, and there is no coast or ocean without a tide. Fundy has an especially large tide, so it appears unusual, but the economic and scientific, social and biological dynamics at play here are also at play for billions of people across our watery planet.

⁓

Several days after my visit to Alma, I perched on a stack of fish traps in Hall's Harbour on the Nova Scotia side of the bay and watched the water drain under the fishing fleet like a long exhale. At low tide, the harbor is completely dry. The boats—*Petrel, Bacchus, Pretty Woman*, and a couple dozen others—settled haphazardly on the gravel bottom. To survive these continuous groundings, Fundy boats are fitted with two keels instead of one. Grounding on a single keel is like trying to balance on one leg; the second keel allows the boat to settle on two

Bay of Fundy fishing boats at high and low tide.

"legs." From my fish-trap seat, which didn't move with the tide, I saw the boats sink below the dock like a setting sun, leaving only their radio antennae visible. At dock's end, a sign reads, "When the tide is out, you have to go looking for it."

The Bay of Fundy Tourism Partnership describes for the many tourists three kinds of tides: vertical, wave, and horizontal. These are not scientific terms, but they're fitting descriptions for tides everywhere. Hall's Harbour is a vertical tide because the shoreline is steep and an onlooker can easily see a dramatic vertical difference between high and low water.

A *wave tide*—commonly known as a *tidal bore*—is a flooding tide that travels upriver as a single wave. Fundy has a wave tide on five rivers—the Herbert, Maccan, Salmon, Shubenacadie, and Petitcodiac—but there are eighty or ninety worldwide, with the largest on the Qiantang River in China and the Amazon in Brazil.

A horizontal tide occurs on gradually sloping shorelines where an ebb tide exposes acres—sometimes miles—of intertidal zone. Sandpipers like the horizontal tides of the upper Bay of Fundy because the mudshrimp are there. "It's the perfect environment for these little shrimp," Hicklin told me, "because they burrow into the mud during low tide and swim at high tide to feed."

I first heard about the sandpipers, mudshrimp, and Peter Hicklin while talking to Rossiter in Alma. I met Hicklin shortly after, eagerly accepting his invitation to return the next year to see the birds. Before I made the trip, he asked that I bring three bottles of red wine from Oregon or Washington and a pair of cut-off jeans. "We can't buy wine from those wineries," he told me over the phone. "You'll need the cut-offs because I'm going to take you out on the mudflats—that's the only way you'll really understand what's happening here."

~

Now I'm back beside Hicklin at Johnson's Mills, having delivered the wine last night. He figures there are 140,000 *peeps*, as ornithologists call them, on the beach in front of us. Some are resting with their necks turned 180 degrees, some are preening, others are nervously pacing. "They're roosting here," Hicklin explains. "They've been foraging for twelve hours without a break, following the tide all the way out and then all the way back in. One bird will swallow about 16,000 mudshrimp during that time. At slack high water they rest for a couple of hours before the cycle begins again."

A semipalmated sandpiper (not to scale).

Rest is a relative word for a bird with a heart rate of eight hundred beats per minute. Like its cousin, the rufous hummingbird, which for its size has the world's fastest metabolism (one thousand beats per minute), the sandpiper's fiery interior never slows, powering long flights and hours of foraging. What little rest a peep gets in Fundy, it gets amid the protection of the flock, which is perpetually alert. The flock watches for predators, including four nesting pairs of peregrine falcons that were reintroduced in the 1980s. These falcons, faithful to one partner and one nest year after year, raise chicks that are eager for a hunting lesson about the time the peeps arrive. An immature falcon is a bit gawky, but a falcon is a falcon, and sandpipers are no match for their speed and ruthless precision. They scatter at the first hint of a falcon's shadow. But the daily losses are a small price for the flock to pay for Fundy's bounty.

Between fueling up and watching for peregrines, the birds take little notice of us. Pressed against one another by the rising tide, they are so close we can study them without binoculars and spotting scopes. I see details of individual birds: delicate dark and light feathers; white chests with fine red and brown stripes; black rumps, legs, and eyes.

Their short, slender black beaks are slightly blunt. They shift and jostle, wing to wing. Their constant murmur—*cher, cher, cher*—is as soothing as the light breeze.

〜

Some scientists argue that the birds land here not by plan or mysterious genetic knowledge but simply from exhaustion in their long flight to Suriname, much as a car coasts to a stop when out of gas. Once here, the logic goes, the birds discover by happenstance a lush tidal environment that answers every need: a safe resting place, a large muddy flat, and millions of tiny shrimp with a fat makeup perfectly matched to their dietary needs. "You can pass this off as coincidence," reflects Peter as we watch the roosting birds, "but I see connections that are far too intricate and interwoven to be mere chance. It's like the puzzle pieces of Fundy evolved together, the shape of one influencing the shape of the others, until they fit only each other, only here."

Hicklin's comment reminds me of a passage in Gary Snyder's book *The Practice of the Wild.* Snyder is a poet whose work challenges our ideas about evolution. What if we looked at evolutionary theory through different eyes? he asks. Instead of seeing species as only competing in a race through time, what if evolution was *cooperative*? What if one thing perceives the need of something else—and calls it into existence?

"I didn't invent that idea," Snyder told me when I met him on *Crusader* several years ago. "Yale ecologist G. Evelyn Hutchinson made the point that evolutionary and ecological processes work together. A habitat might create a condition which will shape the species, or players, that encounter it. And vice versa. When you look at it that way, all sorts of possibilities open up. Huckleberries and salmon call for bears, the clouds of plankton of the North Pacific call for salmon, and salmon call for seals and thus orcas."

This makes me wonder: Did the mudshrimp and birds of Fundy *call* each other? Did the tide, which flooded this region three thousand years ago, call *all* this into existence—the shrimp, the birds, dulsers, fishers like Rossiter?

I'm spellbound as another few thousand birds take to the air, undulating over the water like a long ocean swell. Hicklin stops talking and peers at the flock through binoculars. At sixty, he is fit, wearing jeans and a collared shirt, his salt-and-pepper beard neatly trimmed. He has spent most of his life studying Fundy—hiking the mudflats, gathering mudshrimp samples, banding birds, writing papers, following migrations. When he started his research at age twenty-six, his subject was obscure. No funding was available. Why would anyone want to study such a wasteland? In a sense, his story became the bay's story. The more I learn about him, the more I see that he's as woven into the tide's fabric as the birds and the tiny mudshrimp.

～

Born in Edmundston, New Brunswick, in 1952, Hicklin was raised by his French Canadian mother, who worked at a pulp mill, and his grandmother. When Hicklin was not yet two years old, his British father went for bread and milk one Saturday morning and never returned. The family later learned that he had fled to Australia. "Mom and Grandma spoke only French, so what little English I learned, I got from TV," he says. Determined to be bilingual, he enrolled in Sackville's Mount St. Allison University in 1968, the only fully English-speaking university in New Brunswick at the time. "My first class was introductory English, which I thought was a language class. When the professor announced that we'd start with Shakespeare's *Tempest*, I had no idea what he was talking about. I didn't even know what the word 'tempest' meant."

He graduated in 1973 with a bachelor of arts in contemporary French literature and a minor in biology. "But my real interest was the natural world. After college I got a job as a naturalist at Prince Edward Island National Park. I loved it, but I discovered I knew almost nothing about wildlife. I had a degree in biology but couldn't tell the difference between a sparrow and a blackbird. When I confessed this to a friend, he said, 'You gotta go to Acadia University in Wolfville. That's the only place around here that offers a wildlife degree.'"

Hicklin enrolled the next fall and was first introduced to semipalmated sandpipers while assisting a graduate student who was studying the birds' roosting behavior. "The first day of that project, in late July 1974, we went to Evangeline Beach on the Nova Scotia coast. Migration was just starting. There were five thousand peeps sitting on the beach. I was stunned. 'That's a shitload of birds,' I thought. We knew nothing about them—not one thing."

Soon after this introduction, Hicklin stumbled onto Johnson's Mills, following a rumor of large flocks, and immediately sought university support for a study. His thesis topic was the feeding behavior of semipalmated sandpipers, with a focus on the conversion of fat to energy. He had learned that fat molecules from bugs and worms have a variety of shapes that sometimes fit well with a bird's metabolism—and sometimes not. "It's all about energy conversion. Food goes in as a fatty worm and transforms into energy to fly and forage and breed. Fats that don't fit well take more energy to digest."

From the Evangeline Beach study, Hicklin knew the birds spent half the day on the mudflats. "But what were they doing? They appeared to be foraging, but for what? Where did they come from and where did they go?"

At that time, the mudflats were considered barren and dangerous. No one went out for fear of getting caught in the mud and drowning in a flooding tide, which came in so rapidly it could outrun a person, especially if he or she was knee-deep in mud. There were stories of accidents, even fatalities. For several summers Hicklin collected mud samples at eleven sites, timing his daily visits with the tide, driving the same route, parking his car off the road, donning a T-shirt, old shorts, and a backpack full of empty coffee tins. On the beach, just before trudging into the mud, he removed his shoes and set them above the high-water mark, socks neatly tucked inside. He took samples at mid-tide, about a mile out.

"I got caught once," he says. "I wasn't watching the time, and suddenly I was surrounded by water. My backpack, full of samples, weighed forty or fifty pounds. The whole flat disappeared under cloudy water, so I couldn't see where I was going. I stepped into one of the deep ravines that I usually stay away from because the mud is too soft. You could sink in to your hips and never get out. I stepped in and sank. Water was up to my neck. I had to release the backpack and swim. No one was around to see I was in trouble. From that, I learned to watch the clock and go out only on a dropping tide."

Although Hicklin was finding thousands of *Corophium* in his mud samples, he couldn't link them definitively to the sandpipers. One day he and a fellow researcher drove into Johnson's Mills and found thousands of dead birds. Apparently a very large tide had flooded the beach, forcing the birds to roost on the road, and a car had driven through the flock. "I was crushed, but my friend reminded me of the unusual research opportunity, so we loaded up hundreds of birds in boxes and put them in the lab's freezer. Later, when we split open their stomachs, we found lots of mudshrimp. That's the moment I learned what they were eating."

≈

Hicklin timed our visit to Johnson's Mills not just to see the birds but also to hike over the mudflats in search of mudshrimp. The tide, which comes in fast, slows as it reaches the high point on the beach, then pauses at slack high water for a few minutes. Once it turns, it falls away quickly, at a rate of about ten feet per minute. Within an hour it has vanished, exposing miles of glossy mudflats. With it go the birds, spreading along the interface of water and mud, relentlessly pecking for *Corophium*. "We used to divide into pairs to count pecks," says Hicklin as we slip into the back room of Johnson Mill's Shorebird Interpretive Centre, a small shack staffed by the Nature Conservancy since 2002, to change into our cut-off jeans. "One of us would count pecks out loud and the other would record. An average bird pecks 150 times a minute—about two pecks per second—and gets a mudshrimp every other peck. That's a high success rate, mostly because male mud-shrimp are the last to burrow after the tide recedes, so they're easy pickings."

We cross the pebbly beach and remove our shoes several feet above the high-water mark, just as Hicklin has done hundreds of times before. The mudflats gleam in the afternoon sun, but a thick fog looms in the east. The tide has disappeared and the sandpipers are spread out, busily feeding.

"I used to think of this as a moonscape," Hicklin says as we step into the sticky, purplish mud. "But now I see these mudflats as Fundy's soul." I want him to elaborate, but I'm distracted by cool mud with the consistency of whipped butter oozing between my toes. We sink to our ankles at first, then to our shins and knees. As we leave the shore, I feel a vague but familiar vulnerability, the way I have often felt untying my boat's dock lines before a long journey. Even with an experienced guide like Hicklin, sinking into Fundy's vast, muddy intertidal zone means leaving one world for another. The world below, the one oozing between my toes, belongs to the tide.

～

Every coastline has an intertidal zone. Some are more vertical, some more horizontal, some a combination of the two; some are rocky and some sandy; some are exposed to heavy surf, while others experience no surf. Globally, these zones cover hundreds of thousands of square miles and are as intensely packed with life as a tropical rainforest or coral reef. On most coasts the tide floods twice a day, like a blanket pulled over a bed, and twice a day the blanket is peeled back, expos-ing everything to the sun and air. The continual cycles of drying and wetting make for a highly specialized habitat in which only the clever-est plants and animals can flourish. In *The Log from the Sea of Cortez*,

Steinbeck writes, "The exposed rocks had looked rich with life under the lowering tide, but they were more than that: they were ferocious with life. There was an exuberant fierceness in the littoral here, a vital competition for existence."

To make sense of this complex world, scientists describe four zones of tidal exposure. The first zone—the *splash* zone—is so high it's never under water. Lichen and blue-green algae like it here, where they see only occasional spray from breaking waves or storm surges. The second zone—the *high intertidal*—is covered only during the highest tides. Barnacles, limpets, and snails live here. The *middle intertidal* zone splits its time between wet and dry. Most of the animals that live in the high intertidal zone live in this one too, along with mussels, predatory snails, rockweeds, and red algae. The last zone, the *low intertidal*, is underwater most of the time, going dry only briefly during the lowest tides. This is home to anemones, sea stars, crabs, sea urchins, chitons, and sponges. This low zone is generally the most prized real estate in the intertidal because the residents, unlike their neighbors in the upper zones, don't face the daily risks of drying out in the sun and wind, getting crushed by breaking surf, or being swept away in the turbulence of incoming and outgoing tides. Yet there are many factors, both biological and physical, that influence where an animal will settle. Limpets and barnacles, for instance, risk the dryness of the upper zone for its absence of predators.

In the intertidal zone, every six hours there's a cosmic beat, a rise or fall, a wetting or drying. With each beat the community knows it's time to hide under a rock or crawl out to eat or mate. Opportunistic shore-dwelling animals—bears, deer, raccoons, humans—listen to these beats too. On a downbeat, they know there's a meal to be had and they wander into the exposed intertidal zone, turning rocks and eating clams, seaweed, crabs, and fish. The Tlingit tribes of southeast Alaska have an expression: "When the tide is out, the table is set."

The intertidal zone is such a gritty, soupy, fecund environment that it may have provided the perfect nursery for life's beginnings 3.5

billion years ago. In that era, the moon was much closer to the earth, and the tides may have been several hundred feet high. In places the intertidal zone would have stretched for hundreds of miles, and each receding tide would have left thousands of tide pools, all warming in the sun. Somewhere in that lush environment—not hugely different from the buttery ooze on which I now stand—our ancestors may have sprung to life.

Where and how life began is a matter of speculation, but most hypotheses agree that a dynamic environment, such as the intertidal zone or an undersea hot water vent, coupled with an energy source like the sun, were factors. Inland tide pools are crowded with the amino acids necessary for life's appearance. Every hypothesis asks: How did these ingredients assemble into the first living cell? According to microbiologist Lynn Margulis, this first transition was more miraculous than the later transition from a cell to a full-size human.

Salmon mysteriously return to their riverine birthplace after years of being away at sea. We are still learning how they do this, but scientists agree that smell is one of their navigation tools. What if we were born in a tidepool and our attraction to the sea is a coming home?

~

After we walk into the mudflat, Hicklin stops shin-deep to rest and have a look around. Upbay, we can see the chiseled ochre headlands of Hopewell Cape, and five miles downbay, we spy Grindstone Island, the site of two falcon nests. Directly across the bay is Mary's Point, another popular sandpiper feeding and roosting site. To the northeast—the opposite direction of Mary's Point—the coastline blurs into the estuary of the Petitcodiac River, which flows through the city of Moncton and delivers volumes of inland nutrients to Shepody Bay, mixed and distributed by the tides. From the banks of the Petitcodiac, fields spread for tens of miles, many of them only inches above sea level.

Before the Acadians arrived in the early 1600s, most of this land— 88,000 acres around the bay—belonged to the tide. Thickly popu-

In a process called "tiding," seventeenth- and eighteenth-century Acadian farmers built dikes to drain Fundy's salt marshes and reclaim them for agriculture. The dikes, constructed with an earthen core and grass sod veneer, were called *aboiteaux*.

lated with salt-tolerant grasses and contributing huge quantities of organic matter and nutrients to the tidal mix, salt marshes are among the most biologically rich habitats on earth. "I'll bet the Fundy salt marshes account for about a third of the biological productivity of the upper bay," says Hicklin.

But the Acadians saw the low-lying fields as valuable agricultural land. In a process called "tiding," they built earthen bulkheads called *aboiteaux* that allowed the tide out but not in. Within a few years, rain and snowmelt leached out the remaining salt, leaving layers of rich soil. The newly claimed land was good for growing wheat, vegetables, fruits, and herbs. Fundy hay, famous for its high mineral content, was prized around the world. With so much bounty to be gained by converting tidelands to farmland, Acadians never questioned the value of tiding. A 1760 provincial report captured the prevailing sentiment: "Great quantities of marsh, meadows, and low ground . . . in the Bay of Fundy . . . are spoiled by overflowing of the sea . . . which by industry may be greatly improved."

In the mid-1700s, the English drove out the Acadians, who eventually resettled in the southern United States (ancestors of present-day Cajuns). The English continued the tiding tradition, eventually building tidal barrages or causeways across most of the major rivers. By the early 1900s, they had driven the tide from 70 percent of Fundy's salt marshes. Graham Daborn, a biologist at Acadia University, told me, "It's a matter of conjecture how much more productive these waters would be today if the marshes had been left intact. In all likelihood, the decline of wildlife in the last hundred years was hastened by the loss of so much salt marsh."

From our shin-deep vantage point, Hicklin points out the green-yellow haze on the mud's surface—algae blooming in the open sunlight—and thousands of small snails grazing on it. Here and there, a worm stretches out lazily. Sandpipers are everywhere—preoccupied with feeding, skittering within arm's reach. "Do you see the pinholes?" Hicklin asks. I don't at first, until my focus narrows to a couple of square inches on the slightly crusty, pocked surface. "Each of those is an opening to a *Corophium* burrow," he says as he slips his fingers five or six inches into the mud and pulls up a large handful of it. We both lean in and study the network of veinlike burrows, which are about five inches deep, U-shaped, with two entrances. The *Corophium* keep them from collapsing by mixing a gummy secretion into the sediments lining

A *Corophium volutator* or mudshrimp (not to scale).

the inner walls. The green algae that grow on the mud's surface also secrete a gelatinous substance, all of which helps stabilize the mud.

Hicklin peels open layers with his free finger. "There's an adolescent," he says, pointing, "and there's an adult male." He peels back another layer. "There's a female born this year, and there's one born last year." To see what he sees, I have to narrow my focus even further. With our faces almost cheek to cheek and hovering a couple of inches above the pile of mud in his hand, I begin to see lots of these rice-grain-size creatures. "They're fussy about where they build their burrows," he says. "The mix of sand and mud and silt has to be just right—not too dry and not too soupy; not too coarse, not too fine. I've been here when there's so many *Corophium*—about three thousand a handful—that the flats sound like a bowl of Rice Crispies after the milk is added."

Although I can barely see their shrimplike bodies, much less determine their sex or when they were born, Hicklin points out some of their anatomy. As amphipods (only distantly related to shrimp), their chitinous bodies are divided into many segments, each with a set of legs designed for special tasks like capturing and handling food, absorbing oxygen, crawling, or swimming. Antennae are used to sense the environment and find a mate. Food—bacteria and single-celled diatoms stirred up by the tides—is ground up by a pair of mandibles at the front of the mouth.

Two generations of *Corophium* are produced each year in the upper bay. Eggs are fertilized and developed in the female's pouchlike sac and released as young adults in late May. By early July they're sexually mature and releasing their own young. The *Corophium* population, then, is at its yearly high when the sandpipers arrive in early August. Between predation from fish that follow the tide in and birds that follow the tide out, the *Corophium* population drops off significantly in mid-August. The second-generation survivors overwinter deep in Fundy's mud and emerge the following spring to continue the cycle.

I sink my fingers into the mud and scoop up another pile. This time my eye is better trained and I see hundreds of the tiny translucent creatures. I think of what they must endure: heavy predation from birds and fish in summer, icy winters, massive tides. They compensate for predation by spawning millions of offspring and even benefit from the thinning by sandpipers in late summer, which leaves more room for overwintering groups. They fend off frigid winters by going dormant deep in the mud.

But what about the massive tides? What would be comparable in the human world? A tsunami? A category-five hurricane—twice a day? Mudshrimp would be safer if they stayed permanently in their burrows, but they would starve and never find a mate. They have to come out, and that means coping with the tide. This is not casual: miscalculating the daily tide, even by minutes, would be fatal for these little creatures. If they emerge from their burrows too early or stay out too long, they'll be eaten, cooked in the sun, or raked across the flats with the tide's first rush.

To manage these tsunami-like conditions, *Corophium* have an internal clock that tells them *in advance* when the tide is coming and going. Cockles, oysters, crabs, worms, and a few dozen other intertidal animals have internal clocks too. Fiddler crabs hide under rocks during high tide and emerge during low tide to fossick and mate. Shannies and gobies, small fish that live off the coast of England, sit out intervals of low tide in upper zone tidepools, then roam at high tide.

Even single-celled commuter diatoms have internal tide clocks. They burrow in the sediments during high tide and commute to the sunny surface at low tide to photosynthesize. These microscopic algae are so abundant that they will transform an expansive gray mudflat into a velvety golden-brown bloom. Minutes before being covered by the next flood, their internal clocks warn them it's time to dive for safety, and the mudflats suddenly turn gray again.

⁓

Most of what is known about internal clocks comes from the study of biological rhythms (chronobiology) in land animals. These typically follow daily (solar), monthly (lunar), seasonal, or annual cycles. One of the first scientific reports on biological rhythms was written by Linnaeus in the eighteenth century. In his garden in Uppsala, Sweden, Linnaeus cultivated a "flower clock" that tracked the hours of the day based on different opening and closing patterns of eight flower species. A golden poppy, *Eschscholzia californica*, for example, opened its petals at 9:30 a.m. and closed them at 3:30 p.m., and a field rose, *Rosa arvensis*, opened at 4:30 a.m. and closed at 8:30 p.m. The clock was completed by planting species that opened and closed at intervals in between. Linnaeus's flower clock was so accurate that rumors spread that it would replace the mechanical clock, which was frustratingly unreliable during that era.

Studies of bees in the early twentieth century showed that they could be trained to fly to a food source at the same time each day. Bees trained in France and flown overnight across five time zones to New York maintained their homeland schedule, suggesting they were responding to an internal rather than an external timing mechanism. It has since been discovered that there are hundreds—perhaps thousands—of organisms with internal clocks set to daily rhythms. A sparrow has foraging peaks at sunrise and sunset. The leaves of tobacco and bean plants droop at night, as if sleeping. Lizards change color during the day. The body temperature of a bat rises and falls with day-night cycles. These internal clocks are as fundamental to survival for a land animal as a tide clock is for a *Corophium* or a crab.

These daily rhythms are known as *circadian* (i.e., "about a day"). In controlled laboratory conditions (constant light or dark, constant temperature, etc.), most internal clocks will vary in rate. Humans are a perfect example of this. With our twenty-four-hour wake-sleep pattern, we are classically circadian. Yet when kept in constant light or dark for long periods, we won't all default to a twenty-four-hour wake-sleep cycle, as might be expected. Some will settle into a twenty-two-hour rhythm, some into a twenty-six-hour rhythm, and others between. Because of this phenomenon, scientists use the prefix *circa* for all the primary rhythms: *circadian, circalunar, circannual, circatidal.*

The paradox is that while these internal clocks demonstrate their true "circa" nature in the lab, outside the lab they have the ability to adjust with great precision to specific environmental conditions. A human with a twenty-two-hour rhythm in the laboratory will adjust to a twenty-four-hour solar rhythm outside the lab, like the rest of human society. For a *Corophium*, this adjustment is a matter of life and death. If its clock is not perfectly tuned to the tides, the tiny animal would be mercilessly swept away.

The ability to adjust to changing conditions is one of the internal clock's signature characteristics. This is why a *Corophium* in the Bay of Fundy and its cousin in Normandy can have the same genetic biological clock, each adjusted differently to the conditions under which it lives. If a Fundy *Corophium* were shipped overnight and placed on the mudflats of Normandy, it would continue to express the tide cycle of its home environment, just as the bees did when shipped to New York. But within a few days (if it didn't lose its life in the meantime), its inner clock would synchronize with Normandy tides. The French bees made this same adjustment when shipped to New York, eventually resetting their clocks to New York time. Humans do this, too. When we fly across time zones, we experience jet lag because our biological clocks are still set to home time. Within a few days, they reset to local time. This adjustment is called *synchronization* or *entrainment.*

Biological clocks can lay dormant, too, and be awakened. Fiddler crabs living in the Mediterranean, where the tide range is less than a foot, demonstrate no tidal clock. There's no need for one. If they're transported to a livelier coast, however, a dormant internal clock springs into action within a few days.

To be useful, circadian clocks in land animals must adjust to daily, seasonal, and annual changes in daylight, day length, temperature, and so forth. This adjustment is Herculean enough, but for circatidal

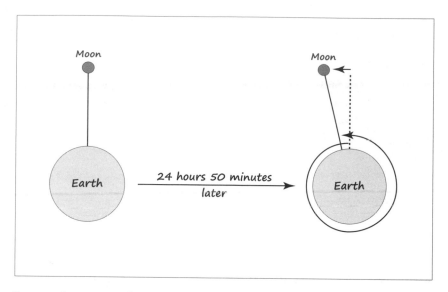

Because the moon is orbiting the same direction as the earth, a lunar day (from moonrise to moonrise) averages fifty minutes longer than a solar day (from sunrise to sunrise).

clocks the task is made fantastically more complex by the fact that there are many tide rhythms. Which tide rhythm should they set the clock to? Some tidal rhythms are so faint a crab or mudshrimp wouldn't notice or bother, but many are not faint at all. The variables are so numerous, in fact, that the same high tide might arrive at one coastal bay at 4:00 p.m. and arrive three hours later at another bay just two miles away.

Most intertidal creatures set their clocks to one or two primary tide rhythms: the lunar day or the spring-neap cycle. Both are lunar cycles, because the moon, although smaller than the sun, is much closer to the earth and exerts about twice the sun's influence on the tide.

The primary tide rhythm of the lunar day, as fisherman Terry Rossiter pointed out, averages twenty-four hours and fifty minutes. It's longer than the solar day because the earth and moon are rotating in the same direction (see figure above). We can picture this if we

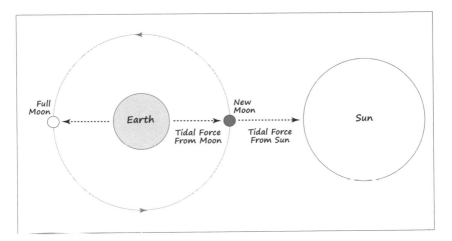

When the earth, moon, and sun are aligned, the sun's and moon's gravitational forces are working together to create larger, or *spring*, tides. These occur at a new or full moon.

imagine ourselves as a child on a merry-go-round. If Mom stands still while watching, she will appear in front of us each time we complete a full circle—say, once a minute. But if Mom walks slowly in the same direction as the merry-go-round, it takes one minute plus a few seconds to catch up with her. In this analogy, the merry-go-round is the earth, stationary Mom is the sun, and walking Mom is the moon.

The spring-neap cycle is the rhythm of the new and full moon. Roughly every two weeks the earth, moon, and sun are aligned (see figure above). When this alignment occurs, the gravitational influ ences of the sun and moon are working together on the earth's oceans, and we see large tides. These are called *spring* tides, from Anglo-Saxon *springan*, "to rise, swell, burst" (no relation to the season). Spring tides bring higher-than-average high water as well as lower-than-average low water. In the weeks between, when the moon is half full (at first and third quarter), the influence of the sun and moon are at odds, and we see smaller tides, or *neap* tides, from Anglo-Saxon *nep*, "lacking or scanty" (see figure on next page).

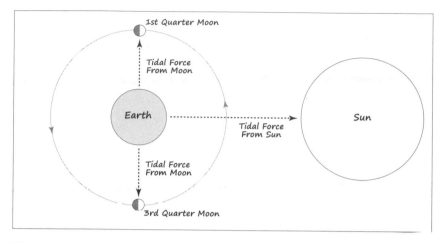

When the moon is in the first or third quarter, the sun's and moon's gravitational forces are working at odds to create smaller, or *neap*, tides.

Fundy's mudshrimp have an internal clock set to the lunar day, which warns them of the tide's whereabouts. But they have another clock too, set to the two-week spring-neap cycle, which tells them the best time to mate and release eggs. If their eggs are released during the three days of highest spring tides, they have the best chance of wide distribution—every species' dream.

The California grunion, a fish the size of a herring, also has an internal pacemaker set to the spring-neap cycle. During spring tides between March and September, it swims from offshore and rides the waves onto the beaches of southern California and northern Baja. Once there, the females wriggle tail-first into the sand and deposit eggs, while the males release sperm. The eggs are fertilized under the sand high on the beach, where they won't be disturbed by high tides again for another two weeks. About that time, the young hatch and ride the spring tides back to the sea.

A recent study by Emily Carrington, a biology professor at the University of Washington's Friday Harbor Labs, confirmed that whelks—a small intertidal snail—have internal clocks set to the spring-neap cycle too. "Whelks hide under rocks and in crevices to stay cool in the mid-intertidal zone," she told me. "They're very slow. It takes them twelve to eighteen hours to eat a barnacle, their favorite food, and about five days to eat a mussel. This is a huge time commitment for such a small animal, and it exposes them to lots of risks—tide changes, drying out, predation. So they have to plan ahead. A crab, on the other hand, can crawl up the intertidal zone, eat a barnacle, and scamper back all in one tide."

Carrington's four-year study, funded by the National Science Foundation, confirmed that whelks fast during neap cycles and binge during spring cycles, using the large incoming tides to boost their shoreward journey and keep them wet while hunting high in the intertidal zone. When they find a good barnacle or mussel, whelks drill a hole in the tough outer skeleton, inject a relaxant, and scrape out the meat with their radula (a raspy tonguelike appendage). Whelks eat nonstop for about four days during spring tide, then inch their way home with a full stomach, only to hide under a rock until the next spring tide.

"We didn't know about the whelk's spring-neap rhythm until now," said Carrington. "Intertidal animals are difficult to study, so we're really just discovering how tightly coupled they are with the tides. I wouldn't be at all surprised if in the next decade or two we find that almost every organism living near the ocean has a tidal rhythm in its genes."

〜

In the mid-1970s, about the time Hicklin began catching sandpipers at Johnson's Mills and recording bill and wing length, weight, sex, and so forth, the Canadian Wildlife Service hired Guy Morrison to band sandpipers in James Bay in northern Ontario, where a large population was known to nest in the spring and summer. He caught birds in mist nets and dyed their breasts and tail feathers bright yellow with nitric acid. A few days later, these bright yellow-marked birds started showing up in upper Fundy—Grande Anse, Mary's Point, Johnson's

Mills. "On the data sheets, there were dots representing populations at Cheyenne Bottoms in Kansas and Quill Lakes in Saskatchewan, but a big blob in the Bay of Fundy. Now we knew our birds were coming from James Bay. But where did they go? The yellow dye wore off before they left Fundy."

Guy Morrison's work was precipitated by interest in developing Fundy's tidal power during the 1970s oil crisis. The state-of-the-art technology for tide-generated power back then was a barrage, or dam, built across a tide stream with gates that allowed the rising tide to fill an upstream holding pond. Once the holding pond was full, the gates would close. As the tide dropped on the seaward side of the barrage, the water in the holding pond was allowed to pass through a turbine on its way back to the sea. In 1965, France had successfully built the first of these barrage-type tide-power stations on La Rance River on the Brittany coast. (With an output of 240 megawatts, it was the world's largest until the 2011 completion of the 254-megawatt Sihwa Lake barrage in South Korea.) With La Rance as a model, Fundy Tidal Power, a consortium of developers, scientists, and engineers hired by the province, proposed three tide-energy barrages in the upper bay, precisely in the areas used by the sandpipers. With this consortium, government funding became available to step up research in several disciplines, including those that would reveal possible ecological impacts.

At age twenty-five, Hicklin was in the right place at the right time. The Canadian Wildlife Service (cws) awarded him a contract to continue his bird studies, which included summer aerial surveys. He was also asked to represent the CWS on scientific panels and at public hearings. "It was a very contentious topic," recalls Hicklin. "The engineers, politicians, and developers were twice my age and incredulous when I voiced my concern for the fate of the mudflats, the *Corophium*, and the birds. 'Son,' they would say, 'there's nothing out there but mud.'"

The growing interest in exploiting tide power made money available not just for bird research but also for studies on fish (cod, herring, and shad, among others), tide science, and sediment dynamics. Very little was known about the tide's interactions with the environment. Scientists knew that river runoff was mixed and distributed by the tide, but not how important this was to the bay's biological diversity and abundance.

"It was a very exciting time. There were lots of community meetings and interviews on the radio. CWS brought me on full-time. They funded my research and asked me to sit on committees. During my high-tide aerial surveys, I counted 560,000 sandpipers in the upper bay *in one day alone*. Morrison's report came out about that time too, suggesting that the total population of North America's semipalmated sandpipers was 1.5 million to 2 million—possibly the world's largest population of shorebirds. Twenty-five percent were flying from James Bay west, down the Pacific coast and stopping in San Francisco Bay. But most of the population were coming through Fundy." Eventually, the tide barrage concept was dismissed, primarily due to new studies showing that blockage of the upper bay would cause unknown, unintended, and unwanted changes in the tide dynamics. The kicker was a report from a sedimentologist who predicted that barrages and all their power-generating equipment would be silted up within a year or two of construction. "He knocked some common sense into all of us," says Hicklin. "If it plugged up, we'd all be screwed. We'd lose our fish, birds, mudshrimp, *and* power."

~

For sixteen years, Hicklin watched sandpipers and collected *Corophium* on Fundy's mudflats. To catch more birds, he modified the standard mist net into a more effective Fundy Pull Net, which tripled the catch. "We banded over fifty thousand birds during that period, but no data came back. The birds seemed to fly south and disappear."

At a 1985 conference, Hicklin met Dutch ornithologist Arie Spaans, who had seen thousands of peeps in Suriname. "Those must be our birds, I thought. But we had no data to show it. Arie Spaans suspected the connection, too, and asked if we had bands he could look for." Spaans, along with colleagues from the World Wildlife Fund, were interested in an international shorebird program that would protect the sandpipers' migratory route. They convinced Dr. George Finney, then director of the Canadian Wildlife Service, to join them in Suriname to discuss the venture.

"After meeting Spaans, I made a bunch of little white flags that could be slipped over a bird's knee. Eight days after the first flag was attached, I got a call from George Finney in Suriname. 'Pete,' he said, 'I just came from the beach. Pete! I saw nine sandpipers with white flags. Our birds are in Suriname!'"

Hicklin went down to Suriname in the fall of 1989 to see the sandpipers and help organize the international shorebird program. Before flying home for Christmas, he went shopping in the capital city of Paramaribo and was hit by a truck. He flew ten feet in the air, slammed against a telephone pole, and dropped to the sidewalk, headfirst. Unconscious and bleeding, he was taken to the hospital. A day later he was flown to Trinidad. His wife, Carrie, and their two sons flew down to care for him. "When I woke," says Hicklin, "I recognized Carrie, but not my own sons."

After months of recovering, Hicklin took a six-month fellowship at Oxford University and then returned to the Fundy bird studies. Even today, having retired from CWS, he continues to sift through data, write papers, and visit the flats. "We've learned a lot in thirty-five years, but there are still important questions to answer," Hicklin says as we retrace our steps back to shore, staying ahead of an incoming tide. "The sandpiper population is shrinking, and we don't know why—it could be a combination of things, but habitat loss and degradation are probably high on the list." I ask if the research he initiated in the 1970s is continuing. "Absolutely," he answers. "Where there were just a few of us back then, now there are many."

Hicklin stops again. His shirt luffs in the breeze as he picks up another handful of mud and gazes down the curves of a silted ravine.

Fine sediment piles like snowdrift on the ravine's edges; its bottom, five feet below, is slick and shiny in the afternoon sun. The scene reminds me of Hicklin's earlier comment about Fundy's soul. Watching him, I'm filled with appreciation for the surprising solitude and beauty of this muddy in-between world. It indeed looks primordial, where life may have begun.

～

Julie Paquet, Hicklin's replacement at CWS, and Diana Hamilton, a biology professor at Sackville's Mount St. Allison University, continue the bird research at Johnson's Mills, and scientists like David Drolet and Myriam Barbeau have picked up the *Corophium* studies.

For the last several summers, Paquet and Hamilton have been banding and tracking birds with radio telemetry. Among other things, they want to know whether the peeps are shortening their yearly stopover in Fundy. "This is important," Hamilton had told me earlier that week when Hicklin introduced us, "because the birds might be reducing their stay due to stresses like increased predation and diminished food supply. If this is the case, they may be leaving without enough calories to make it to Suriname." At the end of our conversation, Hamilton invited me to watch a banding session at Johnson's Mills.

After dropping Hicklin off at his home in Sackville, I drive back to Johnson's Mills. A light rain is falling as I pull up to the banding cottage, a boxy cedar-shaked cabin tucked among maples. It's 6:00 p.m. The tide is flooding, swallowing Hicklin's and my earlier mudprints. High slack, the best time for netting birds, is in two hours.

Inside the one-room cottage, with its small kitchen, round dining table, sofa, and woodstove, Hamilton introduces me to her biology students from Mount St. Allison—Devon, Bill, Mary, Abby, David— and her fourteen-year-old son, Ted. They're all clad in green and yellow raingear, planning for the evening. Julie Paquet, from CWS, sits at the dining table, which doubles as a banding station, and sorts through a red toolbox loaded with gear—bands, orange and green leg flags, needles for drawing blood, a weight scale, calipers for measuring wing and bill length, needle-nosed pliers for crimping the bands.

The goal of the three-year study, a collaboration between CWS and Mount St. Allison University, is to catch as many birds as possible during the first few days of their summer arrival. In addition to the usual data collected during the banding process, forty-five birds will receive a tracking device glued to their inner feathers. Blood will be drawn from a few individuals for diet analysis.

After the banding, several pairs of students will load their cars with camping gear and radio signal receivers and for ten days chase signals across the province. They will hit the known roosting sites at predesignated times to avoid slanting data in favor of tide cycles. Airplane flyovers will be scheduled at high tide every other day. The data, they hope, will answer three questions: How long are the birds staying? Do they move from mudflat to mudflat within the bay? Does their weight gain vary depending on what's available to eat? Hamilton explains that new studies show that while *Corophium* is still the sandpipers' choice food, on some mudflats they're also eating worms and micro-algae.

Ted, Bill, Mary, and Devon are on the sofa, exuberantly recounting the tracking and netting events of the last several days. "We got sixty-five birds in our pull last night," Ted says. "It took about four hours to process them," adds Devon. "And to avoid releasing them after dark, which adds stress, we had to let some go." Hamilton, a youthful mother of two with shoulder-length frizzy brown hair that explodes from under a hair band, is busy testing an electronic tracking device at the kitchen table. "In the last few days," she says, "four hundred birds have been netted, banded, flagged, weighed, and measured. Tonight will be our last night of netting, no matter how many we get."

The team gathers at the water's edge. The sky is gray and heavy; it's an hour before high tide, and eight to ten thousand birds are roosting below the road. The net, made of two twenty-foot poles with fine monofilament stretched between, is unfurled. Ted, tonight's "puller," sets the net near the high tide line and hides in some shrubs up the beach. When the signal is given, he'll pull a string and the net will launch over the roosting birds.

Quietly and slowly, the team herds the birds toward the net. Several large flocks—a thousand or so—are agitated and take to the air. Some return to the beach, but most fly off to the south. By the time the birds are herded into the vicinity of the net, there are very few left. Hamilton nods and the net is pulled. We run with the boxes but are told not to hurry. There are only four birds in the net. Even with so few birds, the team is excited.

In the cottage, a bright work light is hung over the table. Four stations are arranged: one for banding, one for flagging, another for weighing, and the last for measuring the beak and tarsus (between foot and leg). Paquet crimps a band on the first bird and announces, "Adult male, 14738 AYA." Spreading the wings against the light, she adds, "First right primary is broken." Hamilton records the data. AYA, as he will now be known, is quickly passed from one station to the next, finishing the circuit in about two minutes. The team doesn't want to stress the birds any more than necessary.

When all the data-collecting is done, the birds are released on the beach. Hamilton lets the first few go, lifting and opening her palms as carefully as one might throw a light ball to a toddler. She asks if I want to release the last bird. "Show them the water and give them a little lift as you open your hands," she says.

I cup my hands around the small warm body and step toward shore. It's twilight, but I can still make out details. The tide, having reached its peak an hour ago, is slipping away. The flock is spreading out to feed. I open my hands, lift, and the bird flies. The tide is calling.

2 Star of Our Life

A Meditation on Tide History at Mont Saint-Michel

The most admirable thing of all is this union of the ocean with the orbit of the Moon. At every rising and every setting . . . the sea violently covers the coast far and wide . . . and once this same surge has been drawn back it lays the beaches bare . . . as though it is unwittingly drawn up by some breathings of the Moon.

—*The Venerable Bede*, 703 CE

At early evening the sun is sinking into the Atlantic off France's Normandy coast. Its glow, orange and soft in the crisp fall weather, alights on the burgundy leaves of hillside maples and delicate mustard lichens in the stony crevices of Mont Saint-Michel. The spiked silhouette of the eighth-century monastery crowns this tiny granite island like wax dripping from a candle. At 240 feet above sea level, it's the tallest thing on the landscape for tens of miles.

The forty-five-foot tidal range at the Bay of Mont Saint-Michel, which faces the English Channel, is among the world's largest. What makes it unusual is that flood tide completely engulfs the island monastery, save a narrow causeway tenuously straddling the mile between it and the mainland. When the tide recedes, the monastery is left high and dry, like a ship aground.

As I lean on the heavy ramparts of the west terrace outside the abbey's church, I'm drawn by the sand flat's eerie light and textures. It's low tide, the ocean nowhere in sight. In fact, it's more than ten

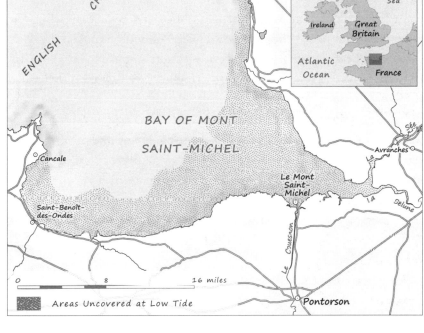

miles away, poised at the bay's head. Soon it will flood, the first wave sweeping across the bay in minutes. For centuries, pilgrims walking from the far corners of Europe to pay homage to Saint Michael have been lost in quicksand or drowned by the deceptively fast-moving flood tide, fabled to be faster than galloping horses. But for now the last of the ebb meanders seaward in shallow, ribboning streams. Knee-high sandbanks border these streams, building in one place and melting in another. Patches of hard, damp sand stretch for miles, their heavy corduroy surface dazzled in sunlight and shadow. The empty bay's otherworldliness is both enticing and forbidding. A guide at the abbey tells me that he walks the flats to shake off his troubles, to forget who he is.

I've planned this trip to coincide with the large spring tides of a full moon. I want to see an incoming tide's "first wave," as the locals call it, and witness the bay's change from empty to full and back again. I want to wander Mont Saint-Michel's abbey and reflect on what our ancestors knew about the moon and tide. Prehistoric coastal dwellers must have been familiar with the daily patterns of each, but when did they first notice a relationship between them? When did they realize that what happens *up there* influences what happens *down here*? How did these notions evolve through ancient times to the Middle Ages, when monks in monasteries like this were collecting and documenting scientific history? It was not a linear or gradual progression but full of fits and starts, braided from the beginning with widely different worldviews, from practical observation to the supernatural and occult, from metaphysics to theology and science. The advance of one perspective often spurred the advance of another, sometimes at odds and sometimes perfectly coupled.

As the sun drops hip-deep in the sea, a flock of gulls takes to the air, squawking. Something is astir. Although I can't detect any changes in the foreground, a molten silver thread wavers on the horizon. Within minutes its ruffled two-foot-high edge stumbles and warps across the flats. I can hear its low rumble above the light breeze and crying gulls. At first, the wave follows the streambeds but soon jumps the banks and spills across everything in sight. In its hurry, it leaves behind a few islands of higher elevation, which are impulsively consumed moments later. The gulls ignore the advancing tide. They fly just ahead, settle on the sand for seconds, then take flight again as it overtakes them.

I'm reminded of a scene in Sir Walter Scott's *Redgauntlet*, which takes place in Scotland's Solway Firth, a horizontal tide almost as dramatic as this. The main character, Darsi Latimer, writes in a letter to his friend, "When I reached the banks of the great estuary, which are here very bare and exposed, the waters had receded from the large and level space." Seeing a horseman spearing salmon, Latimer ventured onto the sand. The horseman began to make for shore, but Latimer lingered. "My steps were arrested by the sound of a horse galloping," he continues, "and as I turned the rider called out to me, in an abrupt manner, 'Soho, brother! You are too late for Bowness tonight—the tide will make presently.' I turned my head and looked at him without answering. . . . 'Are you deaf?' he added—'or are you mad?—or have you a mind for the next world?' So saying, he turned his horse and rode off while I began to walk back towards the Scottish shore, a little alarmed at what I had heard; for the tide advances with such rapidity upon these fatal sands, that well-mounted horsemen lay aside hopes of safety, if they see its white surge advancing while they are yet at a distance from the bank." As Latimer hurried toward shore, the horseman appeared again to warn him of the quicksand. Latimer confessed to being lost, to which the rider replied, "There is no time for prating—get up behind me."

Likewise, brochures today caution tourists not to walk on Mont Saint-Michel's flats without a guide. Besides the risk of quicksand and the tide, lightning is also a danger: anyone walking in the tidal zone is surely the tallest thing on the horizon. "Once the tide hits your feet," warns a local naturalist, "two minutes later it will be at your waist and take you away."

I climb the wide stone stairs to the island's east side, which is still untouched by the incoming tide. The abbey's spired shadow, topped with a sculpture of Saint Michael, stretches out on the clean dry sand. Beyond, the sun's last light burns in the rusty coastal pastures, empty now but by day filled with sheep that thrive on salty, iodine-rich grasses. After its clamorous surge on the west side, the tide seems to stall before finishing its journey around the island. Then it comes, in silence swallowing the abbey's shadow and lapping against its buttresses. At high spring tides, the parking lot and entrance to the small village of Mont Saint-Michel are flooded. During these tides—about four days during new and full moons—the two dozen residents of the village, including the abbey's five brothers and seven sisters, must coordinate their mainland trips with low tides. Delivery schedules,

The Abbey of Mont Saint-Michel at high tide.

like the many trucks that supply the cafés during tourist season, are also governed by the tide.

Looking east, I sight down a fence line and a cluster of trees to mark where I think the moon will rise. As I wait, I offer to take pictures of young couples from Japan, Norway, Australia, and Italy here to see the abbey and the tide. At almost precisely the same instant that the sun disappears in the west, a cool sliver of moon shows itself in the east. It hesitates, as if tangled in the pasture, but soon lifts free and blossoms into a glowing full orb. The tide too has become full, as if this were a planned meeting, one reaching for the other.

〜

If I were standing here sixty thousand years earlier, when human populations first migrated into Europe, I would be gazing at almost exactly the same scene.

As a coastal dweller, I might be huddled around an open fire with a small hunter-gatherer clan, eating shellfish or seaweed gathered during the day's low tide. From experience, I would know that full and new moons mean large tides—low lows and high highs. The sun, with its daily and seasonal cycles of light and warmth, was surely a large part of my experience too. Yet in that distant past it's unlikely I would have noticed a relationship between the sun and the sea. That correlation, which is only half as strong as that of the moon and the sea, wouldn't be discovered until much later in history. For millennia, it was the cycles of the moon that were most mysteriously and conspicuously linked to the earth's waters, especially to the tide's comings and goings. What the moon did up there, its personalities and spiritual qualities, would be forever wedded to the sea.

Peering through ancient eyes, tonight's full moon would tell me that tomorrow's exceptionally low tides would be one of the best foraging days of the month, and I would be ready. But tonight was for celebrating.

I can only guess how our ancestors reveled sixty thousand years ago, but by 1200 BCE, when this island had been established as a sacred Celtic site, my neighbors and I would be dancing, getting married, singing, harvesting crops, or kneeling in prayer. The earliest records tell us that the moon was revered and celebrated as both god and goddess. As god, he was Osiris and Thoth in Egypt, Soma in India, Sin in Mesopotamia. As goddess, she was Artemis, Aphrodite, and Hecate in the Greek and Roman worlds, Isis in Egypt, Inanna-Ishtar throughout Mesopotamia.

The sun and moon were often considered a pair—sister and brother, husband and wife—so in cultures in which the moon was male, the sun was female, and vice versa. Differences in gender, however, were not as distinct as they are today. In *The Moon: Myth and Image*, Jules Cashford writes, "It seems clear that 'the feminine principle,' when referred to the Moon, can take male *or* female form, preventing us from identifying the feminine principle with the human female and the masculine principle with the human male."

Whether viewed as male or female, the moon has always been imbued with feminine principles, perhaps due in part to the close linkage with women's reproductive cycles. In classical Chinese science, the moon was *yin*, meaning soft, yielding, dark, wet, and cold. The sun was *yang*, meaning hard, focused, aggressive, dry, and hot. In Taoist thinking, yin and yang are not opposites but complements—two parts making a whole.

In the grandeur of my view from Mont Saint-Michel, I have no reason to question the limits of the moon's influence. No doubt there are small groups scattered around the world who are worshipping or celebrating this same moon. Its presence, so close and large and luminous, is almost impossible to ignore. Yet sheltered in our homes and cars and cities, few of us know where the moon is and what it's doing. Our encounters are often a surprise, a chance meeting while rounding a bend in the road or taking out the garbage late at night. We might pause and look, feeling a hint of what our ancestors felt so long ago, sharing a moment that connects us through time.

Ancient people observed the moon doing two things at once. First, it rose a little later each night and appeared either more full or less full than the night before. Second, over time (not much time) the moon grew from new to full and then shrank to new again, completing a cycle that repeated over and over. With this, early people witnessed the paradox of a moon that always changes (by day) and always stays the same (by month). The days of the moon's waxing became a time to do things related to growth: conception, birth, planting seeds. The waning days were for tasks related to shrinking and diminishment: curing sickness, gelding animals, harvesting crops.

The few days of full moon were sacred, representing ripeness, wholeness, completion, and potentiality. In contrast, the few days each month when the moon was invisible and "dark," it was thought to have died. Early cultures prayed and offered sacrifice for its return, and when it did—and it *always* did—there was cause for celebration. Horns were blown, dances begun. The new moon, which appeared as

a thin crescent in the dark western sky, was seen as rebirth and res-
urrection, offering hope that humankind could also participate in this
cycle of renewal. In the high drama of the night sky, the moon told the
story of eternity, demonstrating that death is part of life and perhaps
even a necessary step toward rebirth.

In the ascent to heaven after death—a belief shared by numerous
cultures the moon was often the first stop. According to the Hindu
scriptures of the Upanishads, the deceased followed one of two paths,
each passing by the moon. One path returned to earth and the other
united with Brahman at the sun, signifying a reincarnation cycle's end.
"Often," writes Bernt Brunner in *Moon: A Brief History*, "the moon was
imagined as a gate to another world, a mediator between the Earth
and the sun, or a transitional place to an eternal world. Some Buddhist
monasteries have 'moon gates,' thresholds of passage to an altogether
different reality."

The moon was god and goddess, a gate, a symbol of eternity; it
embodied the paradox of change and changelessness. It was also
a timekeeper. Unlike the sun, which could also be used to measure
time (a day), the moon offered a reliable method for reckoning time
over periods of a week or a month. It took seven days (or nights) to
complete each quarter phase and about twenty-nine days to complete
a full cycle. One "moon" became one "month." Once these phases were
identified, they could be used as a point of reference between *what
comes before* and *what comes after*—a context in which to consider the
past, present, and future.

The earliest known evidence of a moon calendar was discovered
in 1967 at the archaeological site of Abri Blanchard in southwestern
France. Amid layers of artifacts, a flat four-inch piece of bone was
unearthed, dating back to 25,000 BCE. On its surface was etched a
serpentine-like series of strokes and notches. According to anthropologist Alexander Marshaak, who first recognized the bone's significance,
the rendering consists of about seventy marks, each depicting the
moon's gradual waxing and waning through two full cycles. Although

The moon's phases in a lunar month. From left: new moon, waxing crescent, first quarter, waxing gibbous, full moon, waning gibbous, last quarter, waning crescent, new moon.

this small bone is the earliest evidence of a moon calendar, Marshaak suggests in his book *Roots of Civilization* that the tradition dates back to the Pleistocene, 300,000–600,000 years ago.

However early humans began using the moon as a primary time-keeper, the practice held sway until the late sixteenth century, when it was displaced in Europe by the Catholic Church's sun-based Gregorian calendar. As Christianity struggled for footing in the early centuries of the current era, the timekeeping properties and mythology associated with the moon were slowly relegated to the profane. The sun grew to symbolize Christ, and the moon's role as fertility goddess shifted to Mary. "Sun-day" became the Sabbath, a day of rest and worship; "Moon-day" became the first day of the workweek. In spite of these large shifts in worldview, many cultures—including pockets within the Christian world—still see the moon as a symbol of spirituality. It's telling that religious holidays such as Easter, Passover, Ramadan, and most Buddhist festivals are timed to the moon. Lunar calendars are still used in China and many Arab countries.

～

At arm's length in front of me, I can stack four fingers between the horizon and the moon. The sun's pinkish light is gone, and so are the other tourists, who are probably eating crepes and walking in the tiny village below. I'm envious, but the solitude up here captures me. I wander through the cloisters and imagine the Benedictine monks of the Middle Ages pacing under the vaulted ceilings. What role did the tide play in their contemplations?

The Benedictines arrived at this church in the tenth century, a few hundred years after it was built. As legend has it, in 708 CE the archangel Michael appeared for three nights in Bishop Aubert's dream, asking him to build a sanctuary. The bishop, who lived in the nearby village of Avranches, dismissed the dream as fantasy. In exasperation, Saint Michael drove his finger into Aubert's skull on the third night. This, as the story goes, finally got the bishop's attention (without killing him), and he ordered that a church be built, which now stands on the island's west side. Over hundreds of years, the Benedictines finished the abbey and quarters, hauling stones by boat from the

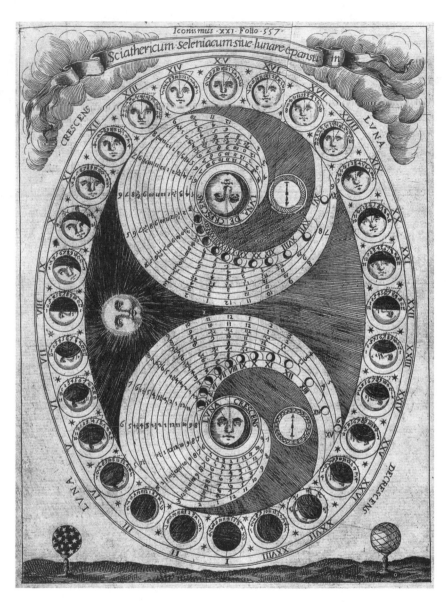

Example of a lunar calendar: "The Selenic Shadowdial or the Process of Lunation," from Athanasius Kircher's *Ars Magna Lucis et Umbrae* (1646), showing the moon in twenty-eight phases.

Chausey Islands at the bay's head. To take advantage of the tides, they sailed out with the ebb and sailed in, fully loaded, with the flood. With several interruptions—for war and a hundred-year stint as a prison— the abbey was occupied by the Benedictines for more than a thousand years. In 1979 Mont Saint-Michel became a UNESCO World Heritage site, and in 2002 it was taken over by the Monastic Communities of Jerusalem. Founded in Paris, the order has communes at nine other locations, including Strasbourg, Brussels, Warsaw, and Montreal.

When I traveled here for the first time in 2012, I wanted to meet the monks who live in the abbey. I wanted to ask them about the tide, but they were elusive among the throngs of tourists, their black robes disappearing in a swirl from one small wooden door to the next. When I left France that year, I began a correspondence with Jessica Martin, one of the abbey guides, who agreed to help me contact the monks. After two years and many delicately worded letters—all translated and choreographed by Jessica—I finally received an invitation to attend a silent lunch, followed by a half-hour of conversation.

I know tomorrow I will meet the monks, but tonight I ponder a let- ter sent earlier from Brother François-Marie. "The abbey," he wrote, "is a jewel set in a beautiful case [the bay]; isn't beauty a language through which God and mankind want to meet?"

If beauty is a language, then Mont Saint-Michel is well spoken tonight. At the western ramparts, where a couple of hours ago I had watched the flood's advancing first wave, I now study the ebb tide's fingers withdrawing in the inky light. I'm struck by the different voices I hear in the tide and the monastery. The abbey speaks of human achievement, built painstakingly in service to the archangel Michael, a guardian and dragon slayer. The monument is an eloquent testament to humanity's insatiable desire for connection and belonging—to God, to nature, to other humans. The stone heft speaks of permanence, a longing for the eternal.

The tide is another kind of eloquence. It's the moon's voice on earth, spoken in perfect synchrony. What the moon does, the tide does.

What the moon *is*, the tide *is*. Shifting by day but repeating a larger pattern, the tide mirrors the moon's paradox of change and change-lessness. In its repetition, it personifies eternity. In its constant flux, it reminds us of what we can't control. It reminds us that our fortresses are impermanent.

The tide's language was surely understood by our earliest ancestors. It's hard to imagine otherwise. If, for example, I were still eating shell-fish around my open fire some sixty thousand years ago, what would I know about the tides? Or perhaps a better question is, what wouldn't I know? Assuming I'm part of a clan that has lived for years among these dangerously large horizontal tides, I would have learned from my elders when low water occurred, how long it lasted, and the best places and times to gather shellfish and seaweed. My survival would depend on it. I would also have learned to take advantage of tidal cur-rents while fishing or traveling in my canoe. In my youth, I might have tested this knowledge and found out the hard way that bucking the tide was not only inefficient but dangerous.

Over hundreds of years—well before the appearance of written history—coastal dwellers would have amassed large amounts of local knowledge. They would have known that there are usually two high tides and two low tides a day, often unequal in height (today called *semidiurnal inequality,* meaning twice daily and unequal). They would have been aware that tides arrive a little later (about fifty minutes) today than yesterday. And they would have noticed that high tides grow and shrink with the moon's phases and that the highest and low-est tides happened during new and full moons (spring tides).

Other patterns would have been obvious too, like how wind blow-ing in the same direction as the tide hurries it along, piling it up earlier and higher than usual. A wind blowing against the tide stalls and flat-tens it. Even a passing storm affects the tide: a low-pressure system means less weight on the sea's surface, so high tides are higher. If a few of these conditions occurred on the same day—a full or new moon along with a heavy storm and strong winds—the tide could be four or five feet higher and arrive an hour or two earlier than usual (what we call *storm surge*).

As early people grew more confident in their seafaring skills, they would have paddled or sailed down the coast, trading goods and stories with other clans. Local knowledge, because it related directly to survival, would have been a highly valued trade item. Even today, when sailors set anchor in unfamiliar waters, it's the locals they trust for the best guidance.

The oldest hard evidence of awareness and practical use of the tide comes from a tidal dock at Kathiawar, in the Gulf of Cambay (today's Gulf of Khambhat), India. Dated between 2500 and 1500 BCE, the dock was an enclosed basin as large as two football fields, with baked-brick walls and a narrow entrance. With the gulf's tidal range of thirty-three feet, ships were required to time their entrances and exits with high water—otherwise they'd go aground. Once the ships were inside the lock, a gate closed to keep them afloat until the next high tide. It was an ingenious solution that helped maintain lively trade between Kathiawar and the earliest known civilizations of Egypt and Mesopotamia, all connected by the Persian Gulf and the Arabian and Red Seas. The locks were built a few hundred years after the Egyptian pyramids, making them one of the ancient world's engineering wonders.

Tides are not mentioned in the Bible, but Moses was likely well armed with local knowledge when he led the Children of Israel out of Egypt in the fifteenth century BCE. Having grown up near the Red Sea (then called the Reed Sea), he would have been aware of the fourteen-day cycles of spring and neap tides. He would have known that during the lowest spring tides, a sandy shoal was exposed that stretched all the way across the Red Sea. In *The Power of the Sea*, Bruce Parker, chief scientist at the National Oceanic and Atmospheric Administration, suggests that Moses planned the great exodus to coincide perfectly with one of these low spring tides. Moses knew, claims Parker, that Pharaoh would send an army of chariots after them. He also knew that

Pharaoh, who lived along the nearly tideless Nile, probably had little experience with fluctuating water levels.

For a successful escape, Parker writes, it was critical for Moses to know "when low tide would occur, how long the sea bottom would remain dry, and when the waters would rush back in." Choosing a low spring tide would have given them longer to cross the dry spit, followed by a higher high water, sure to engulf Pharaoh's pursuing army.

Timing was critical. At the first sight of the dust raised by Pharaoh's army, Moses and the Israelites would know how long they had. The last of them would need to cross the spit just before the tide turned, luring Pharaoh's chariots onto the exposed flats where they would be caught by the advancing tide.

If this is what happened, the story's drama probably grew with each telling until it included pulsing walls of water.

〜

By the time I descend the many steps of Mont Saint-Michel's abbey, even the village below is empty and silent. The moon has almost completely traversed the night sky, leaving dew clinging to café awnings and tile rooftops. Having risen in the east many hours ago, the moon now hovers in the west. The tide is long gone. In a few hours the moon will drop below the western horizon, and just as the last of it disappears, the sun will rise on the opposite horizon. During the days of full moon, the sun and moon are aligned but on opposite sides of the earth (a configuration called *opposition*); during new moon, they are also aligned but on the same side of the earth (*conjunction*). This alignment—either conjunction or opposition—was known to the Greeks as *syzygy* (from *suzugos*, meaning "yoked together"). While in syzygy (see figure on next page), the moon and sun indeed seem tethered. When new, the moon never strays far from the sun, rising and setting on the same horizon at the same time. When full, the moon and sun dance as if attached to either end of a rope, one rising while the other

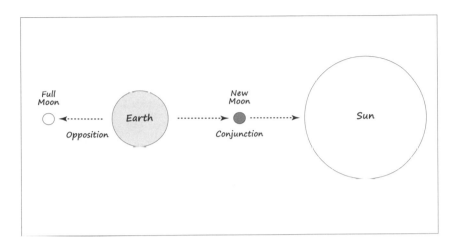

The alignment of earth, moon, and sun is called syzygy. When the moon is full, it's in opposition; when new, it's in conjunction.

sets, one setting while the other rises. We always experience spring tides during syzygy.

As I study the descending moon, it's easy to understand how the earliest people saw a relationship between it and the tides. If I stood here night after night, I would surely connect the moon's position with the tide. In a world full of uncertainty, that kind of knowledge would have given me security and confidence. But it also would have raised more questions. The sea's constant commotion was a sure indication that something big and powerful was at play—but what? Observing a coincidental relationship between the moon and tide was an important first step, but it's another step altogether to speculate that the moon was actually *causing* the tide—that the tide went out *because* the moon set. Once a leap like this was made—and it was made slowly and sporadically—the next step was to figure out how the moon did it. What was the mechanism? Heat? A vortex stirred by the moon's motion? Light? A lunar resident spilling buckets of water? Divine influence?

Until Newton proposed that the moon and sun did it by gravity, answers to these questions were speculative, varying wildly from culture to culture, era to era. The answers also varied due to motivation. Astrologers were inclined to search for divine influences. Natural philosophers ("scientists" before there was a word for them) were often stymied by the idea of "action at a distance." They looked for physical mechanisms such as heat, wind, whirlpools, or compressed air. The earliest suppositions solved the puzzle through myths, many of them nonlunar. A classic Chinese text of the first century BCE speculates that the tides were caused by a sea serpent slithering in and out of its cave. A Malayan legend claims the waters rose and fell with the stirrings of a giant crab. In tribal legends from Canada's British Columbia coast, Flood Tide Woman caused high water simply by lifting her skirt.

The tide's importance to Native peoples of North America is evident in the embellished details of a Tlingit myth about Raven, paraphrased here from Gail Robinson's *Raven the Trickster*.

> Once, long ago, when the world was new, there was Raven. Raven and his people lived near the shore of the "Big Water" before there were tides. They enjoyed the clams that occasionally washed up on the beach, but as their population grew, there was less food. The Great Spirit spoke to Raven in a dream: "I have seen that you and your people are suffering. . . . There lies at the end of the world, at the edge of the Big Water, a cave. In this cave sits an old woman who holds the tide line across her lap. This controls the ebb and flow of the water. . . . If you can get her to let go of the line, the water will fall and the people will be able to get some good things to eat below the surface. This will not be easy, Raven, for the woman holds the line very tightly."
>
> So Raven flew and flew to reach the cave at the end of the world. Always the trickster, Raven pretended to be eating a clam until the old woman took notice and asked, "Raven, where did you get those clams?" Raven ignored her until she got close enough for him to kick sand in

her eyes. Blinded, the woman let go of the line and the tide went out. Realizing what had happened, Raven couldn't wait to get home and enjoy a good meal. For many days, he and his people feasted, but soon the creatures exposed by the low tide started to dry up and die. The people begged Raven for help, so he flew back to the cave.

The old woman was still clearing the sand from her eyes. "Is that you, Raven? You tricked me!" she said. "Yes," said Raven. "I did trick you. I wanted to get the good things to eat when the tide went out, but now the creatures below the ocean are dying. If I help you, will you help us by letting go of the tide line from time to time, so we can eat the creatures and they can stay alive?" The old woman agreed, so Raven cleared her eyes and sat her back in the cave with the tide line across her lap. From time to time, she lets the line go, and that's how the tides began.

Raven, forever the trickster, figured prominently in other Northwest Coast origin stories. He pecked open the box of daylight and placed the sun in the sky; he dropped a clam on the beach, splitting it wide and hatching the first humans. In the myth of the tide line, through his greed he teaches the value of moderation, compromise, and balance.

The ancient Greeks had their tricksters too, but they saw the "Big Water" as a living being and the tides, its breath. Plato believed the earth was a large animal and the tides were the sloshing of its inner fluids. A few hundred years after Plato, the Roman geographer Pomponius Mela wrote: "It is moreover not quite understood whether the world causes that [tide] by its own breathing . . . or whether there are certain caves sunk below the surface where the . . . waters reside and whence they rise up copiously again, or whether the moon is the explanation of such great movements."

The tide as evidence of a living earth persisted for a long time. At the beginning of the Renaissance—about two thousand years after Plato—Leonardo da Vinci jotted in his notebook: "As man has within him a pool of blood wherein the lungs as he breathes expand and contract, so the body of the earth has its ocean, which also rises and falls every six hours with the breathing of the world." Da Vinci was so taken by this idea that he tried to calculate the size of the earth's lung.

I'm taken by this idea too. Even with the benefit of modern astronomy and science, one cannot spend time on or near the ocean without feeling that it's alive. No wonder we project human qualities onto it. Yet isn't this part of the beauty it speaks—that it breathes as we breathe? As I study the earliest imaginings about the tide's cause, I remember something from the architect and inventor Buckminster Fuller: "When I am working on a problem, I never think about beauty. I only think about how to solve the problem. But if the solution is not beautiful, I know it's wrong."

The Hindus worked on tide problems too, and their solutions were every bit as beautiful as da Vinci's. The tides appear in the sacred texts of the Vedas, the Puranas, and the Upanishads. The twelfth-century BCE Visnu Purana claims that heat is the cause: "Like the water in a cauldron, which in consequence of its combination with heat expands, so the waters of the oceans swell with the increase of the Moon." Taking this concept a little further, the thirteenth-century Persian writer Zakariya Qazvini wrote: "As for the flow of certain seas at the time of the rising of the moon, it is supposed that at the bottom of such seas there are solid rocks . . . and that when the moon rises over the surface of such a sea, its penetrating rays reach these rocks . . . and are reflected back thence; and the waters are heated and rarefied and seek an ampler space and roll in waves toward the seashore."

≈

These first tide theories grew out of cultures that lived near large tides. Ironically, the Hellenistic world, which was largely responsible for documenting these theories and passing them on to the West, borders a sea with almost no tide. With the exception of the upper Adriatic (near present-day Venice), the Mediterranean's largest tides rarely exceed a few inches; Naples's spring tide, for instance, is six inches. Because of this, we hear nothing about the tides from the Greeks and

Romans until Herodotus's first mention of them in the fifth century
BCE. Even then, it's the tides of the Red Sea he cites, not those of the
Mediterranean. News of tidal phenomena trickled back as other trav-
elers ventured out.

Pytheas of Marseilles was the first Greek to sail past the "Pillars
of Hercules" (Straits of Gibraltar) and return with stories from the
British Isles and beyond. His writings, which reported exaggerated
tides of 120 feet, perished but were summarized by later philosophers.
About the same time, Alexander the Great, during his ten-year campaign (336–326 BCE) to conquer "the ends of the world and the Great
Outer Sea," met with surprisingly large tides in the Indian Ocean that
almost ruined him. He died in Babylon, never having made it home,
but his army returned to Greece with sea tales that piqued the interest
of natural philosophers.

What also came back from Babylon with Alexander's army, as if
lodged in the crevices of their leather shoes, was astrology, the sister science of astronomy. Developed by the Chaldeans and spread
throughout Mesopotamia, astrology's core belief was that human
destiny could be charted by the movement of planets across the sky.
Classical Greek philosophers including Socrates, Plato, and Aristotle
were familiar with the Chaldean cosmology, but it wasn't until after
their time that this worldview flourished in the West. By the beginning
of the current era, Greece had developed it to a high degree, passing
it to Rome, the Arab world, and eventually Europe, where it thrived
throughout the Middle Ages.

Lest we doubt astrology's role in the formative years of science,
it's helpful to remember that for the better part of history, almost all
natural philosophers and astronomers—Pliny, Strabo, Posidonius,
Seleucus, Ptolemy, and Albumasar, to name only a few—were also
astrologers. Even as late as the Renaissance, some of the greatest
thinkers—Copernicus, Brahe, Kepler, Galileo—did mathematics by
day and wrote horoscopes for kings and queens by night. Newton,
the father of the clockwork universe, practiced alchemy, astrology's

cousin, and almost poisoned himself while conducting late-night experiments.

Perhaps the most fertile patch of Greek soil in which astrology's seeds took root was Aristotle's "system of the world." In his view, which dominated Western thinking until the seventeenth century's scientific revolution, the moon sat at the boundary between the terrestrial and heavenly spheres. The terrestrial sphere—between earth and moon—was imperfect, mutable, and composed of four elements: water, earth, air, and fire. The heavenly sphere was perfect, immutable, and composed of a fifth element, *aither* (Latin for "pure air"). The outermost sphere was the domain of the *Primum Mobile*—the Primary or Unmoved Mover—which was itself motionless but caused all other motion. (Jews, Christians, and Muslims later replaced this entity with God.)

Astrologers had no qualms with Aristotle's system. In their search for specific and meaningful relationships between the celestial and earthly spheres, astrologers wanted to make things more personal. They did that by putting the moon in charge of soil and water and the sun in charge of air and fire. The sun dried things out; the moon made them moist. Coupled with everything wet, the moon was now in charge of dew, rivers, blood, and the sea. Because the sun had no business with anything watery, to suggest that it played a role in the tides was heresy.

Until Newton's day, almost all significant inquiries about the tides were viewed through this lens. When the moon waxed, earth's moisture also waxed: sap ran, blood expanded, breasts swelled with milk. Human temperament was subject to these influences too. Hippocrates, the father of Western medicine who lived a generation before Aristotle, proposed that human disposition was dominated by the four bodily fluids, or *humours:* blood, yellow and black bile, and phlegm. These humours changed in volume with the moon's phases, causing mood swings ("lunacy" was the moon's influence on the humours of the mind).

These were not extraordinary claims for the era, but many of them were hard to prove. How do you demonstrate beyond a doubt, for example, the advantage of being born during a waxing moon or that melancholy is related to a waning moon? How do you prove that what happens *up there* affects what happens *down here*? Astrologers reasoned that if a cause-and-effect relationship could be demonstrated in one case, it would affirm divine celestial influence in all cases.

They didn't need to look far. An obvious relationship existed between lunar cycles and women's reproductive cycles—so obvious, in fact, that the two were considered to have sympathetic natures. Ovulation was the waxing moon, bleeding was the waning moon. The reproductive cycle was a metaphorical tide within a woman's body.

But perhaps a closer link between *up there* and *down here* was the sloshing of the ocean tides, which could be seen mirroring the moon's every phase. Could there be more convincing proof of divine influence? And if the moon was capable of creating such a stir in the earth's oceans, what else could it do? What about her celestial brothers and sisters—the planets and stars? How were they influencing human destiny?

As recognition of the moon-tide partnership grew, it's not surprising that the moon's spiritual qualities were projected onto the ocean. The lunar gods and goddesses—dark and feminine—became gods and goddesses of the ocean. The tide's coming and going reflected the moon's coming and going, each inseparable and personified in the other.

Like a waxing moon, then, a flooding tide was a time of growth and an ebbing tide a time of decay and diminishment. James Frazer writes in *The Golden Bough* that coastal dwellers were apt "to trace a subtle relation, a secret harmony, between its tides and the life of man, of animals, and of plants. In the flowing tide they see not merely a symbol, but a cause of exuberance, of prosperity . . . while in the ebbing tide they discern a real agent as well as a melancholy emblem of failure, of weakness, and of death."

Animals die only during an ebb tide, wrote Aristotle. Two thousand years later, Charles Dickens wrote in *David Copperfield:* "People can't die along the coast . . . except when the tide's pretty well nigh out." If a sickly person survives the ebb, "he'll hold his own till past the flood, and go out with the next tide."

~

On the day of my interview with the monks, rain and wind cloak Mont Saint-Michel. I leave my hotel early, seeking a few uncrowded hours at the abbey. When I reach the parking lot, it's flooded by the sea. The narrow footbridge arches gently and promisingly but disappears mid-span into the turbid water. No one is around, so I take off my shoes and pants to wade. The water is cold but soothing as it rises to my shins, then knees. When it hits my waist, I hold my clothes above my shoulders and wonder how much deeper it will get. Did someone forget to let go of her tide line? Luckily, the water gets no deeper, and I emerge dry from the waist up at the abbey's arched entrance.

For religious pilgrims who walked for weeks or months across Europe to pay homage to Saint Michael, this last two-mile stretch was probably the most treacherous. Yet they came by the thousands, especially during the Middle Ages when Mont Saint-Michel was as popular a pilgrimage as Rome, Jerusalem, and Santiago de Compostela in northern Spain. Travel-weary, and with their destination looming within reach, most pilgrims were not prepared for this last challenge. They had no experience with tides, and in that era there were no printed tables listing the times of high and low water. If they had a few extra shillings, they could hire a guide to take them across. But most could not afford it and had to risk the crossing on their own.

Of the many legendary stories of loss, one tells of a pregnant woman who wanted to pray to Saint Michael for the safe birth of her child. Her husband and attendants took her across the sand flats, but on their return the tide overtook them. In a panic, the woman gave birth. The husband and attendants tried to carry her and the child to safety but could not. "In tears," the legend goes, "they hastened to leave, because the sea was already in front of them." The woman, left behind, appealed to Saint Michael in desperation: "In the name of God, don't let me drown here, die in this sea." Her prayers were answered and the little patch of sand where she lay remained dry. "During all this time," reported the woman, "I felt there was around me a curtain, much more white than the snow on a branch; it looked like a wall: the sea could not get over it." To commemorate the miracle, she named her child Peril.

Today, during July spring tides, a yearly pilgrimage organized by the diocese attracts hundreds of pilgrims to walk the five miles from Genets to the Mont. Genets was the village of departure for pilgrims in the Middle Ages.

I put on my pants and, with some hours to spare, look for a place out of the rain. I settle at an open café, grateful to be under a roof, and read through my notes on some of the pilgrimages undertaken by ancient sailors and philosophers. They weren't seeking salvation, but they were pilgrims nonetheless, embarking on unknown seas and foreign soil, risking whether they would make it home again.

Many of the books written by early Greek explorers were lost, and consequently most of what we know about that period was collected and summarized by just a few writers, notably the geographer Strabo (Lucius Seius Strabo, 63 BCE–24 CE) and the natural philosopher Pliny (Gaius Plinius Secundus, 23–79 CE).

Strabo, in his seventeen-volume *Geographica*, cites the Rhodes philosopher Posidonius, who studied the tides for a month on the coast of Spain. According to Strabo (and, later, Pliny), Posidonius's most noteworthy discovery was that the Gauls had been observing tides long enough to recognize not just daily and monthly, but also yearly, patterns. They knew about the fourteen-day rhythms of spring and neap tides, but they told Posidonius that spring tides grow incrementally larger and smaller over cycles of months, even years.

The record isn't clear, but if the Gauls had indeed identified these longer-term tidal rhythms, then they had come upon a new discovery.

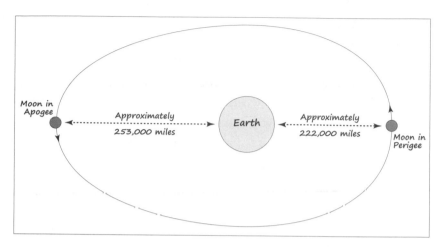

The moon's orbit is elliptical, at times closer (perigee) and at times farther (apogee) from the earth.

Without the help of modern astronomy, however, they could not have known that what they witnessed on the beach was orchestrated by the moon's lopsided orbit around the earth. Because of this, the moon is sometimes closer to the earth, exerting more gravitational influence, and sometimes farther away, exerting less (see figure above). When the moon is closest, in *perigee*, it's 222,000 miles away, and the tidal force is 22 percent greater than average. When it's farthest away, in *apogee*, it's 253,000 miles from us, and the force is 16 percent less than average. This is a gradual change, hardly discernible from day to day but very much discernible at the extremes. To an observer on earth, even the size of the moon changes with this cycle. Tonight's moon over the Bay of Mont Saint-Michel is in apogee. As bold as it is, it would appear a quarter larger if I came back during perigee, twenty-five days from now.

When the moon is full or new *and* in perigee, the tides are extra large. This happens about four times a year (twice during full moon and twice during new moon) and is called a *perigean syzygy* or a *perigean spring*. At modern-day Cadiz, where Posidonius first learned of this, a spring tide during apogee is a little over eight feet. During perigee, it's about eleven feet, a difference of three feet.

If the Gauls detected a yearly fluctuation, then it would most likely have been due to the eccentricity of the earth's orbit. Just as the moon's orbit around the earth is lopsided (elliptical), so is the earth's orbit around the sun. With close observation, the Gauls might have noticed that spring tides creep farther up the beach in January when we're closest to the sun (*perihelion*) and recede in July when we're farthest away (*aphelion*). This is about a 5 percent variation, equal to about a foot of difference on a coast with twenty-foot tides.

One way or another, the Gauls must have been sophisticated observers. What would have motivated them to study the tide over such long periods, especially when the differences were hard to measure and too subtle to be useful in food gathering or boat navigation? Perhaps the observations helped with timekeeping. Whatever the reason, their discoveries cracked open a door that revealed new dimensions of mystery and complexity—and wonder. Where they previously thought only two or three tidal cycles were at play, now they could see four or five or six, stretching from days and weeks to months, even years. Without the laws of planetary motion, which came almost two thousand years later, they couldn't have guessed that there were actually far more than six tidal cycles. By the twentieth century, more than four hundred were identified, some as short as six hours (a partial lunar cycle) and some as long as 25,800 years (the earth's wobble, also known as *precession of the equinoxes*). Each one of these cycles has a tidal effect.

Like Strabo, Pliny the Elder, born in Como, Italy, in 23 CE, wrote extensively about tides, capturing most of what was then known. Together with other natural philosophers, he discussed the work of mathematician Seleucus of Babylon (ca. 150 BCE), who supported the heliocentric theory, which posited that the earth revolved around the sun. The theory, first proposed by Aristarchus in the fourth century BCE, was eventually revived by Copernicus in the sixteenth century and widely accepted by the late seventeenth century. Seleucus may also have been the first to suggest a plausible mechanism whereby the moon causes the tide. If the earth's soupy atmosphere extended to the moon, he reasoned, then the moon's motion would rub against it and cause waves that were felt in the earth's oceans.

What Seleucus's theory omits, as did all early theories, is a reason for the tide's repeating twice-daily pattern. One tide could be accounted for: the one that happened when the moon was above the horizon and visible. The second tide, which happened when the moon was below the horizon, was much harder to reconcile. How could the moon be causing a tide when it wasn't even visible?

By Pliny's time, the moon's role in exciting the earth's oceans was well established. In *Natural History*, he writes:

> The moon is not unjustly regarded as the star of our life. This it is that replenishes the earth; when she approaches it, she fills all bodies, while, when she recedes, she empties them. . . . During sleep, it draws up the accumulated torpor into the head . . . and relaxes all things by its moistening spirit.

Pliny's *Natural History*, translated from Greek to Latin in the Middle Ages, remained an authoritative survey on tides well into the Renaissance. A few other significant contributions were made during this time, most notably by the Venerable Bede (673–735), a scholar and Benedictine monk, and the Muslim astrologer Albumasar (787–886).

The Venerable Bede, a student of astronomical timekeeping and often considered the father of English history, spent his life at the Jarrow Abbey on England's North Sea coast, about three hundred miles north of Mont Saint-Michel. The tides at Jarrow are almost as large as those at the Bay of Mont Saint-Michel. In the monastic tradition of the Middle Ages, Bede's writings were largely a synthesis of previous scholars, particularly Posidonius, Seleucus, and Pliny. He added a few discoveries, the most original being the observation that tides behave differently at each location and that they progress in a wavelike fashion down the British coast. He wrote: "For we who inhabit the various coasts of the British Sea know that wherever one tide begins to swell another simultaneously recedes. . . . Not only that, but on one and the same shoreline those who live to the north of me will see every sea tide both begin and end much earlier than I do, while indeed those to the south will see it much later. In a given region the moon always maintains whatever bond of union with the sea it once formed." This last line may be the first written reference to something coastal dwellers knew empirically for a long time: the tide's arrival is foreshadowed by the position of the moon in the night sky. Locals may not have known, however, that this relationship was different on every coast.

Albumasar lived a couple of generations after Bede, but like the Benedictine, his work was translated into Latin and survived well into the Renaissance. In his lengthy treatise, he devotes five chapters to tide theory, most of which is a summary of earlier thinking. He lists eight tide characteristics, among them the effect of wind, the unequal heights of the two daily highs and lows (first described by Seleucus), and the influence of the moon's positions and phases. Although he says the sun "assists the moon," he seems reluctant to admit the sun's role. As a devoted astrologer, Albumasar would have verged on heresy if he admitted that the sun influenced anything but fire and air. Albumasar's description of the moon's varying distance from the earth (perigee and apogee) and its tide influence is the first to appear in the written record, indicating progress in tracking the motions of heavenly bodies in the early years of the current era.

≈

As I hike the abbey's 273 steps to meet the monks, dozens of spirited schoolchildren with backpacks and sweatshirts and ponytails prance effortlessly past. They've come by busload on a field trip from neighboring communities, accompanied by worried teachers and chaperones who are huffing, like me, to keep up. When I get to the top, Jessica is waiting. She tells me that each year four million people visit Mont Saint-Michel, but only a quarter of them hike all the way up to the abbey. "If I were a nun," she says, "climbing the stairs every day would be part of my spiritual dedication to God. But since I'm not a nun, I have to find another reason. In my six years here, I've seen this place in almost every kind of mood. It's never the same—the light, the tide, the shape of the flats, the weather. I think it's the lesson of change that keeps me coming back."

Brother François-Marie meets us outside the abbey church, and we follow him through a narrow wooden door, which opens into what used to be the abbots' quarters but now houses a kitchen and small dining hall. As the door's iron latch clunks shut, I'm struck by the warmth and brightness inside. Unlike the larger-than-life edifice on the other side of the door, which sometimes feels cold and heavy, this interior is human-scale. The ceilings are low and whitewashed. Open shelves are crowded with nuts, cheeses, coffee, canned peaches, and ravioli. A basket brims with fresh baguettes. Brother Sebastian, whose turn it is to cook, tends a large iron pot on the stove, thickening the air with aromas of hot butter and steamed rice. The other four monks, along with several visiting pilgrims, carry dishes, sort cheeses, stoke the woodstove. All this is done with a sense of informality, yet with seamless and exacting attention. And no one utters a word. The only sound—and it is sound enough—is the incidental tapping, shuffling, clinking, washing, and stirring as the tasks are carried out.

With a nod we are invited to remove our coats and follow Brother François-Marie into the clean, stone-floored eating room. A wood-stove sits against the south wall; two gabled alcoves face east, each with floor-to-ceiling stained glass. We sit on wooden stools around a horseshoe-shaped oak table. Plates of greens and thin-sliced salami are passed around, then rice, and finally cheese and grapes. Jessica and I quickly learn the gestures to indicate "Do you want some?" and "Yes, please" or "No, thank you." The silent meal, like its preparation, has a brisk, fluid pace. In fifteen minutes, the monks are back on their feet. Everyone carries his own dishes to the kitchen. Someone washes, two of us dry, and two others sort and stack. I'm reminded of the month I spent in a Zen monastery in the hills of Okiyama, Japan, where the daily tasks were carried out with this same clean efficiency. "Everything's like breathing," explained the roshi; "one thing follows another, like fish scales or shingles laid on a roof, with no gaps or interruptions to distract the mind."

In ten minutes the kitchen is immaculate. François-Marie invites us back into the dining hall. Filling our delicate cups with coffee, he is the first to break the silence: "Sugar?"

When all four monks are gathered—François-Marie, Sebastian, Laurent-Nicolas, and Jean-Gabriel—we pull our stools in close and lean toward one another across the table, Jessica and I on one side and the four of them on the other. "I never thought about this subject before receiving your questions," begins Laurent-Nicolas, "but as someone who has consecrated his life to God, I think the main question is, why did God choose this place for a sanctuary when he could have chosen any other place? I think it's because of the tide! Pilgrims came here to seek salvation with God. They knew nothing about the sea. They were terrified, as if it was a different world. But they were ready to die while walking across the tide."

"Now they come not for God, but for the natural wonder," adds François-Marie. "Someone once said of this place that you can hear it listening to you. When I listen back, the silence and beauty sometimes speak louder to the soul than hours of discussion."

"It's a mystery, really, why millions of people come here, precisely here, as if attracted," reflects Jean-Gabriel, who stands behind the other brothers. "We know why we're here—to dedicate our life to God—but what about all these other people? Lots of them come because it's the largest tide in Europe. During spring tides, the hotels are full. During neap tides they're almost empty. The tourists come and go with the tide—just *like* the tide."

Jean-Gabriel pauses and brings his hand to his forehead, as if trying to coax out the next thought. "I think the tides remind people of why they are here on earth. They can be moved by the tide, and then there is only one more step to being moved by God."

"Yes," adds François-Marie. "A woman at Mass said this place is the most beautiful in the world. It shows that people are moved. This kind of experience can be a passage for people who don't believe. In the Catholic faith, we have a credo: 'I believe in God the Creator of the sky, heaven, and earth.' People may not be aware of it, but coming here is a meeting with their Creator. The beauty of this place, its strong tides, is a passage: they meet their Father. They see God through the tide."

The conversation quiets as we take our last sips of coffee. I am filled—but not with coffee or lunch. The monks have spent longer with me than planned, and I don't wish to overstay my welcome. Jessica, who has been translating, expresses our thanks as we make our way to the door.

Outside, the day is still dark and wet. The wind has come up. Fat copper gutters gush bucketloads of water onto the stone steps. Puddling, the water spills in sheets from one step to the next. Pigeons coo from hidden crevices. We climb back to the abbey, where Jessica must return to work. "Now I'll have to rethink the reason I climb these stairs every day," she says.

"And I'll have to rethink the reasons I travel so far to see these tides," I reply. We say our good-byes, for now, and she slips through another small, heavy door to greet the next group of tourists.

Before descending, I walk once more to the west terrace. In the driving wet, I'm the only person out here. I squint to see across the bay where clouds mass on the horizon and shroud the tiny island of Tombelaine, Mont Saint-Michel's sister. The tide is ebbing. Where sunlight played on the sand a few days ago, today the light is muted, sullen.

The idea that we're drawn to these places, perhaps unwittingly, as part of a spiritual journey, will linger in my mind for a long time. If it's intimacy and connection with the natural world that we crave—that I crave—then standing in the wind, the sun, the tide, is as close to that as I will know. As I make my way down toward the village, soaked but content, I remember Brother Sebastian's parting words: "Tourists are pilgrims who don't know they are pilgrims."

3 Silver Dragon

China's Qiantang River Tidal Bore

This is the strangest and most wonderful sight under heaven.

— *Mei Sheng, second century* BCE

∼ I wake before dawn on the last day of China's Bore-Watching Festival, when large tides occur during daylight hours. The air is warm, thickened with smells of wastewater and pesticides in the dusty streets of Yanguan. Hidden below the jetty a few hundred yards away, the Qiantang River—one of China's largest—slinks toward the sea like a drowsy dragon. It sleeps now but will awaken with the next tide. During festival days, when the tides are extraordinarily large due to a full moon, the dragon rears up into a twenty-five-foot wave—the largest bore in the world—and terrorizes everything in its path.

Slipping through the gates where festival tickets are taken by day, I make my way to the riverbank. On the jetty, a woman practices tai chi in the moonlight; a man smokes a cigarette on the bleachers. Candy wrappers and Styrofoam cups carpet the seawall top. Nearby, four shirtless men play cards under a bare incandescent bulb. In a few hours this stretch of river will be packed with people.

More than a million tourists come each year to the Bore-Watching Festival, some to pay spiritual homage, others to celebrate the holiday with family and friends. They come from Yanguan, from Shanghai, seventy miles northeast, or from Hangzhou, forty miles upriver.

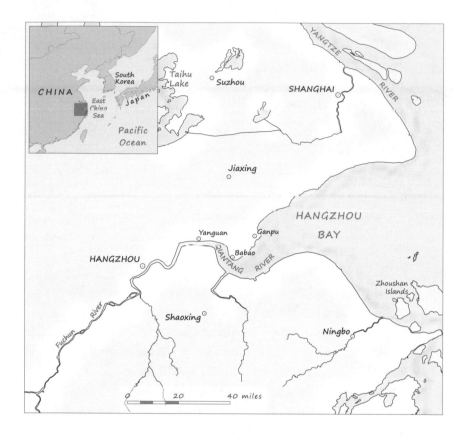

Some travel thousands of miles by train from the western provinces of Qinghai or Sichuan. Vendors sleeping under tables will rise to unfurl colorful umbrellas and sell eggs boiled in tea, dried fish, Wrigley's chewing gum. Men on bicycles will wander the crowd selling melons and sugarcane. Everyone waits for the tidal bore.

Twenty miles downriver, where the Qiantang meets Hangzhou Bay, the tide rises and falls exactly as it does on most coasts around the world. But inside the river's entrance, it becomes a monster. Squeezed by the river's funnel-shaped sides and bottom, the tide erupts into an avalanche of whitewater. Known as Yin Long (Silver Dragon), it spits and jerks through the Qiantang's bends and shallows. The dragon finally collapses a hundred miles upriver, only to rear up again at the next flood tide.

The Chinese have been haunted by this monster every day—twice a day—for at least 2,500 years. In earlier times, when the Qiantang and Yangtze shared a mouth, the monster favored the Yangtze. When the rivers separated some two thousand years ago, the dragon shifted to the Qiantang. Throughout the region, it has inspired poems and stories, scientific analysis, spiritual revelation, and engineering feats. "The waters in the Crooked River will roll on rising waves as high as mountains . . . gathering up a force that threatens to engulf the sun and sky," wrote fourth-century BCE poet Zhuangzi.

The Silver Dragon has also wrought unimaginable destruction. Not infrequently, weather and tidal conditions conspire to amplify the bore's size, and it jumps the dikes, destroying acres of low-lying fields. Peasant farmers, who raise silkworms, tend livestock, and till the rich riverside soil, are often caught unaware and drown. But the preponderance of drownings occurs among bore-watching enthusiasts whose curiosity lures them too close. Over the years, thousands have perished, like crumbs swept from a table.

Yet I too want to get closer. I first learned of the Qiantang's unusual tide in the early 1990s while reading George Darwin's book *Tides and Kindred Phenomena,* written in 1898. Darwin, son of the famous evolutionist, wrote: "There are in the estuaries of many rivers broad flats of mud or sand which are nearly dry at low water, and in such situations the tide not infrequently rises with such great rapidity that the wave assumes the form of a wall of water. . . . Notwithstanding the striking nature of the phenomenon, very little has been published on the subject."

A hundred years later, there was still little written on the topic. In the early 1990s, I searched the Internet for "bore," only to be redirected to "boar," a male swine. "Tide" led to laundry detergent. I sent letters and faxes to China that were never answered, so I finally gave up. For all I knew, the dragon had died for good. Years later I ran across a

Japanese film of the Qiantang's bore and immediately began planning a trip to see it.

Now, on the jetty's edge, I wonder how close I'll get. I don't have a plan, but I didn't come all this way to stand behind a fence. Knowing the dragon-bore will not reach Yanguan for several hours, I climb down the seawall and wander the mudflats. The ebbing tide is swift but silent, exposing neat rows of boulders and rotted piles. A light breeze, cooled by the river, dries my sweat. In this river-bottom world twenty-five feet below the jetty, I am almost completely alone. I lean against the seawall's base, mesmerized. Standing waves loom for seconds, then sink into the chocolaty water. Whirlpools cast off dimples of tension. A man squats chest-deep in the river, his hands feeling for crab in the rocky crevices. Green-netted weirs hang haphazardly on bamboo poles, ready to snare eels swept in by the next bore; the catch will be shipped live to restaurants like A Shan in Shanghai, where I ate a few nights ago.

As the hours pass, I settle on the mudflats and decide to stay as long as I dare. I'm overtaken by a longing to lose myself in the river. If I felt it were safe, I would strip my clothes and swim, let the dragon take me under, push me around, spit me out. I imagine the many intrepid Chinese who have swum in this river, many of them drawn to it as I am. In fact, from the tenth to the sixteenth centuries, when the bore was largest in Hangzhou, thousands came to see it and to watch a swimming competition in which the participants tested their luck by jumping into the river, carrying colored flags as they tumbled in the violent water. They were called *nong'chao'er*, "tide players." The poet Sun Chengzhoung wrote: "Don't marry the tide player. The tide doesn't keep its word." So many tide players drowned, the practice was eventually outlawed.

I decide to stay out of the river. I choose a firm spot on the mudflat and practice scaling the seawall. The time it takes to reach safety will determine how close I can let the bore get to me. The first attempt takes ten seconds, which may not be quick enough. Rehearsing every step, I eventually reduce my time to eight seconds, then seven. That might be fast enough. I'm inside a bend in the seawall, within view of the 120-foot-high, eleventh-century Bhoda Pagoda. The bore will be hidden until it rounds a corner a hundred yards away. I do the math: At twenty miles per hour, the dragon devours twenty-nine feet per second. As soon as I see it round the corner, I'll have ten seconds to escape.

I don't know if I'll have the nerve to stay in its path that long.

～

Like others who come to the Bore-Watching Festival, I journeyed far. I arrived several days ago, and although my twenty-hour flight from Seattle may not be worthy of spiritual merit, it was arduous. At the Shanghai airport, I meet Huang Ying, who holds up a sign among the crowds: "Jona Than." Short and thin, with large brown eyes and long bangs, Ying is a twenty-year-old engineering student at Hohai University in Nanjing. Through a mutual acquaintance, she agreed to be my interpreter. Months earlier, I had asked her to help me find people knowledgeable about the Qiantang's folklore, spiritual history, science, and engineering. I did my own research in advance, but I hoped my week with Ying would lead to interviews and experiences I could never discover traveling on my own in China.

To my surprise, she has arranged a meeting for this very first night. I throw my bag in a cab's trunk, and we're soon weaving through streets that, even after dark, are crowded with people. Household goods and bicycle parts are traded on sidewalk blankets as young couples walk hand in hand. Our driver takes a quick jog into a darkened residential neighborhood, stopping outside the home of Chen Jiyu.

With a warm greeting, Dr. Chen invites us to sit in his small living room. Director of the Institute of Estuarine and Coastal Research at East China Normal University, Dr. Chen sits opposite us, a handsome man in his eighties, then crosses his legs and quietly ponders my questions. He sweeps his thick white hair back with a stroke of his hand.

The Qiantang bore traveling upriver, as seen from above.

"I've been studying the Qiantang River for more than fifty years," he says. "It's been my life." He has published more scientific papers than he can remember, most of them on the Qiantang's unique hydrology. "The gift of this ferocious river," he muses, "is that it's impossible to ignore. It demands our attention. Out of necessity, it has made experts of us."

The Qiantang bore has been an ominous presence at least since the beginning of written history. For survival, ancient people were attentive to its comings and goings. Before the bore's patterns were understood, riverside villages constructed of straw were frequently leveled. After observing the seasonal changes in the bore's course, villagers learned to build their huts on the river's east side during one season and move them to the west side during another. This intimate knowledge evolved into the world's first tide table, carved in stone sometime during the first millennium CE. For each phase of the moon (new through full and back again), the table predicted the bore's arrival time and height. Larger bores were shown during spring tides and smaller ones during neap tides. The table also showed that the largest bores arrived a couple of days after full moon, indicating an awareness of a delayed effect between what happens in the heavens and what happens in the ocean (this delay was later called "the age of the tide"). A printed copy of the stone tide table circulated in 1056 CE, a full two centuries in advance of the first tide table (for the "London Brigg") developed and celebrated by the Western world.

Dr. Chen is too modest to point out that until modern times China's tide knowledge was likely more advanced than the Western world's, but he's clearly proud of the Qiantang bore. There may be more than a hundred bores worldwide, he explains. Most are small and intermittent. The Amazon's famous Pororoca ("Great Roar" in the indigenous Tupe language) reaches heights of twelve feet and is felt five hundred miles upriver. The Colorado had a fifteen-foot bore, called the Burro, but it's now hardly evident due to agricultural water control and consequent silting. The Mascaret on the lower Seine in France and the Eagre on the Severn and Trent in England were once notoriously destructive, but these too were diminished by silting caused by industrial development. Bores are also found in the Bay of Fundy, Turnagain and Knik Arms in Alaska, the Solway Firth in Scotland, the Hooghly River in India, and the Mekong River in Vietnam.

There is no general rule about the shape of a bore's wave, nor its consistency. Some bores advance silently in a series of unbroken swells; others break in midchannel or farther upstream. Still others form only during the highest tides of the month or year. The Qiantang bore is unusual not only because it's the highest and most destructive, but because it forms on *every* high tide, twice a day, every day of the year—and has done so for more than two millennia.

All rivers with tidal bores have two things in common. The first is a funnel-shaped river mouth with a shallow, gently sloping bottom. The second is a large tide at the river's entrance. When this large tide—in the Qiantang's case, twenty-six feet—encounters the shallowing river mouth, the energy shifts dramatically, the way a jet can exceed the speed of sound. The jet's energy shift is heralded by a sonic boom; the tide's is heralded by a monster wave and the sound of galloping horses. It's been said that a tidal bore is a sonic boom traveling upriver.

Dr. Chen shuffles into the next room, his slippers dragging on the hardwood floor, and returns to refill our teacups and hand us each an exquisitely wrapped package. Ying whispers that he has brought us moon cakes, a pastry shared with guests and family during the national holiday known as the Moon Cake Festival, which coincides with the Bore-Watching Festival. Moon cakes are filled with cream or dried fruit and decorated with the Chinese character for "roundness" and "reunion," symbolizing the full moon of autumn and the family's hope for harmony and closeness.

"If we go further back in time," Dr. Chen says as he returns to his seat, "we find different views about the bore's origin." A fifth-century BCE story describes how a young warrior named Wu Zixu was assassinated by King Fu Chhai. His body was cooked in a cauldron, sewn into a leather sack, and thrown into the river. Restless and angry, Wu Zixu's spirit avenged itself by raising the ocean tide and overwhelming the city of Hangzhou, the capital of the empire. King Fu Chhai fought back by firing arrows into the great tidal wave's belly, which, as one account claims, "appeared in the offing like a silver rainbow, then wild horses flying, then a thousand angry dragons." Over the centuries, descriptions of the bore echoed these evocative metaphors as observers tried to put into words its magnitude and power. Desperate to appease the spirit of Wu Zixu—and convinced it could not be extinguished by force—the king finally made offerings of food and flowers and built pagodas along the riverbank.

Wu Zixu was more than a restless spirit. He was a real warrior who lived in the fifth century BCE. Although details of his life have been embellished, the essence of his story is true, woven through thousands of years of Chinese art and literature. He is mentioned in a second-century BCE poem by Mei Sheng entitled "Qi Fa" (Seven Stimuli). The poem is a conversation between an ailing prince and a visitor who offers a cure: "Although one may have a chronic, long-term illness, [viewing the bore] still will straighten a hunchback, raise up the lame, give sight to the blind and hearing to the deaf."

When the prince asks to hear more, the visitor continues: "When [the bore] first rises, a torrent pours forth like a flock of swooping white egrets. . . . All alone it gallops. . . . Roaring, crashing, vast, boundless . . . It bursts into anger, is blocked, starts to foam. . . . Such magical things are eerie and baffling, and cannot be described completely. . . . This is the strangest and [most wonderful] sight [under heaven]."

Wu Zixu's spirit is still alive and well on the Qiantang: the Bore-Watching Festival commemorates his birthday. Pagodas line the river at Yanguan, the most popular viewing spot, and flowers and food are still offered to assuage the angry spirit. At the town's Sea God Temple, pilgrims pray to Wu Zixu's twenty-five-foot-tall statue. During festival days, red carpets adorn the temple courtyard and cauldrons smoke with burning candles and incense.

The Silver Dragon stimulated a spiritual awakening for fourteenth-century folk hero Lu Zheshin. As described in the classic *Outlaws of the Marsh*, Zheshin, a Robin Hood–like hero, was resting in Hangzhou's Temple of Six Harmonies after a battle. He was startled by the sound of "thousands of cavalry galloping together." Fearing the enemy, he grabbed his weapon and dashed outdoors. A resident monk stopped him, explaining it was only the tidal bore. The words stirred Zheshin's memory of his master's hymn: "When you hear the tide, Pass. When you see the tide, Away." He dropped his weapon, changed into clean garments, and sat in the meditation hall. He wrote:

In my life I never cultivated goodness,
Relishing only murder and arson.
Suddenly my golden shackles have been opened;
Here my jade locks have been pulled asunder.
Alas! Old Faithful of the Qiantang River has come;
Now I finally realize that I am myself!

Zheshin, a warrior-monk who never read the scriptures and knew only slaughter, recognized the bore's calling, and there he died with the passing of the bore, ending his life an enlightened Buddha. "The hymn is significant," Richard McBride, professor of Buddhist studies at St. Louis's Washington University, told me over the phone. "It plays on the seminal idea of buddhahood being 'thus come and thus gone.' What comes and goes, is here and gone, and follows the natural course of things more than the tide?"

The stories of Wu Zixu and Lu Zheshin underscore how thoroughly the Silver Dragon has captured and shaped the Chinese psyche. Perhaps the bore is both gift and curse, inspiring a richly layered religious and cultural identity as well as fear of destruction and death.

I worry that I am keeping Dr. Chen awake, but he insists on continuing in his easy, meditative manner. He speaks in Chinese but often corrects Ying's English interpretation. Well into the night, I ask questions; Ying interprets; Dr. Chen gently clarifies. Before leaving I ask if he'd felt a desire to get closer to the bore. "Oh, yes," he answered without hesitation. "But not in the same way that you do. As a scientist, I want to get closer so I can measure it. Poets like you want to get closer to feel its power and beauty."

Surely not everyone wants to get close to something so dangerous. Of the millions who visit the river each year, most are satisfied to watch from a distance. But what about the hundreds who put themselves at risk by venturing beyond the fenced-off areas? Dr. Chen speculates that these people are following an innocent curiosity, unaware of the danger. "For some it's a game," he said, "with deadly consequences."

≈

On my second day, Ying and I hire a taxi to follow the Silver Dragon's path upriver. From Shanghai, we skirt the Yangtze River delta for several miles, its chalky brown waters congested with industrial boat traffic. Our taxi progresses in fits and starts, accelerating at full speed into open pockets and then stopping suddenly, waiting and watching like a predator for the next opening. I envy the stream of rickshaws and bicycles that flow effortlessly by—many stacked tall with building materials, fishing gear, or caged chickens.

A couple of hours later, we pull over at the coastal village of Ganpu, and I get my first glimpse of Hangzhou Bay at the Qiantang's mouth. Here, at the interface of river and sea, the bay is broad—twenty miles across—and ruffled by a light breeze. Above the bay, the Huangshan Mountains, where the river originates, rise like an apparition on the western horizon. From those mountains, the river drops into the lowlands of Taihu Plain, carving broad, snaky ribbons before empty-ing into the East China Sea. With a drainage area of thirty thousand square miles, the Qiantang is the largest river in Zhejiang Province. The combined sediments flushed to the sea by the Yellow River, the Yangtze, and the Qiantang create an extraordinarily rich marine environment that stretches thousands of miles down the coast. The Zhoushan Islands, just offshore of the Qiantang delta, are the most productive fishing grounds in China.

As I look out from Ganpu at the river, its mouth is ominously calm. There's not a boat in sight. This area's *Sailing Directions*, an essential guide for conscientious sailors, warns that during spring tides the bore is "most formidable and navigation is impossible." Vessels over 150 tons simply don't venture inland of the river's mouth. With or without the *Sailing Directions*, local sailors never question the bore's authority. In 1888 Captain William Usborne Moore had to learn the hard way.

A hydrographer for the British Admiralty, Captain Moore sailed the

HMS *Rambler* into Hangzhou Bay and anchored just offshore of where I am standing. In his journal, Moore writes, "I therefore considered to spend two or three days in the Hang-chau estuary, and endeavor to find out the height, speed, origin, and general character of the Bore of the Tsien-tang-kiang; and at Full Moon in September, before visiting Shanghai for coal and leave, I proceeded in Her Majesty's ship under my command."

On the morning of September 20, Captain Moore set out with three gunboats, each loaded with instruments, tide poles, and a week's supply of provisions. Their intent was to ride a flooding tide to the river's mouth and then continue to Haining City (present-day Yanguan), another twenty miles upriver. Due to extraordinarily bad timing, all three boats went aground by midmorning, and for the next three and a half days, the crew and their boats suffered one calamity after another. Captain Moore writes of the first incident: "Here I touched the ground, and being unable to stop the boat from going on to the bank, let go the anchor . . . the boat swung violently round as far as the keel would let her; and, signals notwithstanding, the boats astern, being unable to stop their way, rapidly came up to the same spot. . . . This was the commencement of the southern branch of the Bore. We had started an hour too soon."

Captain Moore noted—perhaps enviously—how local sailors used "shelter platforms" to navigate the river. Immediately after the bore passed, more than thirty junks ran upstream with the after-rush. Before the tide reversed, they tucked inside one of these small shelters at the river's edge and allowed themselves to go aground as the tide dropped. A bulkhead extending into the river from each platform deflected the next incoming bore and, once it passed, the junks were able to safely float again and set sail upstream. Using this strategy, sailors took three days to travel from the river's mouth to Hangzhou. The strategy is still used today, although the platforms are in poor repair. Junks are often suspended on davits secured to the top of the seawall, allowing the bore to pass underneath.

Junks resting on "shelter platforms" awaiting the passage of the Qiantang bore. From Captain Moore's journal, 1888.

After struggling for several days to keep his boats afloat and crew safe, Captain Moore was forced to return to his ship, which had also experienced difficulties with the large tides and unpredictable currents. Not to be deterred from his mission, Moore traveled overland to Haining City. He posted several men along the riverbank to observe the bore, recording its height and character as it passed—once during the day and once at night. An observer reported: "The Bore passed with a loud roar, as one continuous cascade from bank to bank. . . . Over the channel, the crest of the cascade was ten to eleven feet high; but there was [*sic*] many waves carried along in the rip on its back which were three or four feet higher. For several minutes after it had passed, the bosom of the flood was violently convulsed, heaving to and fro, from one bank to the other like the surface of the ocean after it has sloshed over a sinking vessel, now and then leaving a depression in the centre and washing up to within six feet of the sea wall."

For all his interest in the bore, Captain Moore was dismissive of local knowledge. "They can give no reasonable explanation of the flood coming in this particular manner," he writes of the river sailors, "but all emphatically declare it to be a bad Fung-shui (sic). It is not surprising that they should make such a phenomenon the object of superstition and the theme of romance, devoted as they are to . . . the occult influence of air and water." As I gaze out toward the spot where Moore had anchored, I wonder how the locals, tucked safely behind their shelter platforms, might have characterized the captain's tidal knowledge as they watched him wrestle with the incoming bore.

~

Before traveling on, Ying and I stop for lunch. As she reads me the menu, I feel increasingly embarrassed by my provincial diet. I've tasted walrus, fish eyes, and even chicken's feet but draw the line when it comes to large beetles and duck's chin. Ying, on the other hand, has a voracious appetite for all things Chinese. Half my size, she routinely eats me under the table. On this occasion, to my bowl of fried rice, she downs half a chicken, duck's blood soup, rice, and one medium-sized frog. She tells me that when she was a child and didn't finish all her food at the family table, her mother threatened to make her eat American food for a week.

At Babao, ten miles upriver from Ganpu, we join a throng of spectators. They crowd the river's edge, which is lined with colorful umbrellas to provide shade from the intense sun. Barbecue smoke hangs in the air; young adults sit atop the jetty, sucking raw sugarcane and nibbling moon cakes. I'm surprised by the number of women in evening dress—skirts, stockings, high heels, makeup—who awkwardly make their way in the mudflats exposed by low tide.

The rumble of the tidal bore, which forms ten miles downriver at nearly the exact spot where the HMS *Rambler* was anchored 120 years ago, can be heard forty-five minutes before it arrives at Babao. A hush sweeps through the crowd at the first sound of it. Soon after the bore is formed, a six-mile-long sandbar splits it in two. The north branch is seven to nine feet tall and from this distance appears like a milky ribbon across the water. The south branch shifts course and accelerates as it disappears behind the bar. When it emerges twenty minutes later, its trajectory is more northerly, or cross-river: directly at the crowd.

Until now, both branches of the bore have tumbled along at fifteen miles per hour without much event. But when the north and south branches collide, each with the momentum of a twenty-five-foot tide behind it, a thunderous clap riffles across the river, and a mushroom of white water spits thirty feet into the air. Swaths of brackish foam are ripped clean, thrown skyward, and vanish into mist. The confrontation is so clamorous I think surely the bore has exhausted itself, but part of each wave manages to pass through the maelstrom and continue on its original path. The point of collision darts across the river like a closing zipper, slamming headlong into the Babao Dike.

Usually the dike holds its ground while the dragon throws its tantrum. The bore doubles over and storms about with high standing waves. Unable (this time) to breach the dike, it gathers itself and hurries upriver. The whole explosive spectacle is over in a few minutes. In the face of such fury, it seems miraculous that the dikes hold up at all.

"Building and maintaining these seawalls," says Han Zengcui, who has joined us at Babao, "is one of the three largest engineering feats in China's history." The other two, he explains, are the Great Wall, which is more than 10,000 miles long, and the Grand Canal (the world's largest artificial river), which connects the 1,100 miles between Beijing and Hangzhou.

Dr. Han is a consultant and professor at Zhejiang Institute of Hydraulics and Estuary in Hangzhou. At sixty-five, he has just returned from a two-week horseback tour of Inner Mongolia and is presently serving on a government commission charged with planning for the Qiantang's future. Today, because vehicle traffic is restricted during the Bore-Watching Festival, he has arranged an official government escort to take us to Yanguan, about fifteen miles upriver, to get the full effect of the Silver Dragon. As we continue our drive upriver, Dr. Han explains that seawall building has been a part of river culture since ancient times. Initially, dikes were made of pressed earth, which were easily washed away, as were the crops and homes of farmers brave enough to inhabit the lowlands. By the ninth century CE, earthen dikes were replaced with bamboo cribs filled with stone.

One of the many challenges then—as now—was how to protect the foot of the dike from the incessant scouring caused by tidal currents. The tenth-century solution was an apron constructed of two rows of wooden pilings lashed together and driven deep into the dike's base. When these failed, boulders the size of small automobiles were added, but these too were easily undermined and carried away by the bore. The oldest dike still serving its purpose is the fish-scale wall, which gets its name from the layering of hewn stone bonded together with a mixture of sand, lime, and boiled rice. These walls, which first appeared during the Ming Dynasty (1368–1644), are still holding their own along much of the northern bank.

The last five hundred years have seen only minor changes in seawall design. Concrete blocks have replaced hewn stone; gradually sloping walls have replaced vertical ones; steel pilings are driven alongside those made of wood. Mortar did not replace boiled rice as a bonding agent until 1940.

During the Qing Dynasty (1644–1911), officials in designated areas were responsible for maintaining stretches of seawall. Where damage occurred, the official and his family were held responsible. If found negligent, they were often put to death. Today the Qiantang River Administration employs full-time crews to face the many challenges of dike maintenance. Because the river is shallow, for example, boats carrying heavy equipment such as cranes and pile drivers cannot get close enough to the dike without going aground. The dike's top isn't sturdy enough to support heavy equipment either. Consequently, most new construction and repairs are undertaken by hand. These challenges are exacerbated by the fact that fieldwork is limited each day to the few hours of low tide.

"The future of the bore is controversial," says Dr. Han, as we arrive at Yanguan. "Some think it's a miracle and others want to diminish it or use it in some way." Proposed river-training projects ("fathering," as he calls it) include a hydroelectric dam, an underwater spur dike spanning the river mouth, and a manmade island dividing Hangzhou Bay in two (and thus reducing the tidal range). "The coin always has two sides," he says as he drops us off at our hotel. "Many of us have spent our lives trying to improve the situation in the Qiantang estuary. We cannot deny the magnificence of the phenomenon, but on the other hand it is disastrous to the common people who live in the coastal area. The only thing that stands between them and the bore is an ancient and failing dike."

~

Though that ancient and failing dike stands up to most tides, as I witnessed at Babao, the battle isn't won decisively. If a spring tide coincides with one of several typhoons that hit this shore during the summer monsoon season, the scale tips in favor of the bore. With heavy winds and low pressure, a typhoon can add five to seven feet to the tide, with catastrophic results.

The *Zhejiang Disaster Record*, which is housed in most major libraries, lists hundreds of such events. An entry from 775 reads, "In one night of July, typhoon and high tide hit Hangzhou. Five thousand houses were flooded; thousands of boats were swallowed; over four hundred people died." And in 1472: "Typhoon met spring tide. In Hangzhou, ramparts collapsed, farmhouses floated down river, people and animals drowned. In Yanguan, meters-high water overtopped the river's surface. Hundreds of thousands of people drowned."

Records are inconsistent, but at least twelve of these "perfect" storms have occurred in the last seventy years. Typhoon Sinlaku was such a storm. On September 8, 2002, it crashed into the coast with ninety-mile-per-hour winds and extremely low barometric pressure. The moon was new and in perigee, giving birth to one of the largest tides the sun-moon partnership can deliver: a perigean syzygy. The near-perfect coincidence of typhoon and tide raised the largest bore in fifty years. The *Shanghai Star* reported the next day that Hangzhou had registered record-high tides of more than thirty-three feet:

> The storm tide . . . swept away more than one hundred tide watchers in Hangzhou, Zhejiang Province, leaving four seriously injured. The strength of the tide was claimed to be second only to that of a tsunami in its destructive capacity, according to the *Hangzhou Daily*. At 2:50 p.m., with a huge sound, the raging Qiantang River rushed over the dam by several meters, deluging tide watchers at the Hangzhou Jiuxi Water Station as well as passing vehicles. . . . As it approached, the tide rose suddenly and reached the level of the railway bridge, more than ten meters [thirty feet] high. . . . "Cars flew like leaves when the tide struck," a witness said. An ambulance was sandwiched among smashed cars with its siren wailing continuously.

Although many were injured during Typhoon Sinlaku, astonishingly no lives were lost. Drownings, however, are an accepted part of life along the river.

Just a month before my visit, Liu Taiquan, a thirty-four-year-old resident of Hangzhou, lost his only son to a large August tide. After learning that Taiquan wanted to publicize his story to prevent further losses, Ying arranged a meeting. I wasn't sure I wanted to sit with him, for fear of the emotional intensity. I'm a father too, and on that day I was missing my ten-year-old son who was a world away in Washington State.

Near midnight on a hot and sticky evening, Ying and I wait for Taiquan at the entrance to the Fish Distribution Market in Hangzhou, where he and his family have a home and sell ice for a living. When he finally arrives, he shakes my hand and immediately turns back into the dingy and fetid market, motioning us to follow. We barely keep up with his brisk, deliberate pace, nearly losing him several times as he weaves through acres of bustling commerce. Carts, some motorized

A large bore, coincident with an offshore storm, overwhelms riverside viewers.

and some pedaled, brim with live seafood as they race between nar-row alleys. Bare bulbs cast dim light that glistens on slick fish and eel. Smoke from dry ice creeps along the floor; the stench of fermented food, mixed with fish smells, is nauseating. Netted bundles of crab and turtle are jammed next to crates of frog and whitefish. There is no room to walk.

Without a glance back to see if we're still following, Taiquan climbs a steep, narrow stairway to a small room where he, his wife, and their eight-year-old daughter live. We sit down to talk but, unsatisfied with the location, he moves us twice more before settling into a private sleeping room off the main market. It's stuffy but quiet. Taiquan is husky, with short bristly hair and thick arms. He wears brown slacks, soft shoes, and a beige button-down shirt. He tells his story slowly and carefully, punctuated with emotional pauses.

"My sixteen-year-old son, Liu Tao, was on summer holiday and joined my cousin and me to deliver ice. On the way, we saw kids playing on the spur dike at Qi Bo. My son wanted to join them, so we stopped on our way back. My cousin and I took off our clothes and jumped in the river, while my son agreed to stay with the clothes and wallets. It was a little after four in the afternoon. Just ten minutes later I heard Liu Tao yelling that the tide was coming. Before I could swim to shore, the tide surged around the corner and drove me to the bottom. I was pinned there, trying to grab an edge of the spur dike, but the wave pushed me up, then down again.

"I'm a strong swimmer, having grown up on the Yangtze, but I could do nothing against the bore. I finally grabbed the hull of a boat that was anchored nearby and pulled myself aboard. I heard voices yelling for help and threw a life ring to four teenagers clinging to a log. As I pulled them in, I was thinking of my son. I grabbed a young girl by the arm as she drifted by. She was unconscious, but as soon as I got her in the boat she gagged and woke up. I called out for my son . . . went back in the water with a life ring to search for him. I looked for almost an hour. The rescue teams showed up and pulled me out. I was forbidden to go back in.

"I stayed on the bank all night, watching the river for signs of my son. His body surfaced at noon the next day, in front of us."

Taiquan buries his head in his arms and asks to be left alone. I am torn up inside and want to stay with him, but respecting his request, Ying and I quietly leave the room and make our way through the market, oblivious now to the bustle. We hail a cab and ride back to our hotel in silence. It's well after midnight when I nod goodnight to Ying and close the door to my room. I undress and stand in the shower, hoping to cleanse the raw emotional encounter as easily as I wash away the fish market's sticky grime. In the night I wake thinking of my son, and weep for Liu Taiquan and his family.

⥲

At dawn of the last day of the Bore-Watching Festival, before climbing down the jetty and counting the seconds as the bore arrived, I come upon a dozen elderly women chanting softly, burning incense, and lighting red candles on the riverbank. For the previous twenty-four hours they sat in silence or recited sutras in the Buddhist temple nearby. I watch as the women plant burning candles and incense in hard soil and then amble to the edge of the cobbled dike to offer respect to the muddy river. Kneeling, palms clasped, they bow repeatedly, pressing lips to the dewy grass.

The women finish their yearly ritual as the first hint of sunlight spills over the East China Sea. Some will return to their families in the nearby village of Yanguan; others will board westbound trains. I stay at the river's edge for a while, contemplating the women's prayers and still thinking of Liu Taiquan. Eventually I descend to the muddy river bottom, where I practice climbing the wall and resume my counting.

Before long, the river's spell is pierced by a bellow. Although I can see only an eerie expanse of mudflats, my limbs begin trembling. A sudden change in pressure stirs a chill wind. I have traveled a long way to experience the Dragon's power, but it was never my plan to get *this* close. Part of me wants to dash for the wall, but the impulse is hushed by a roiling wall of white water lunging around the corner. With just two seconds—exactly—to take it in, I exhale and count: one one-thousand . . .

I see that the bore is not a single wall of water but a compressed and stacked series of waves. A six- or seven-foot breaker leads the charge, cutting the river's calm surface; a three- or four-foot second wave trails by a few yards; a third breaker somersaults on top of the heap.

Two one-thousand . . . As a sailor and surfer, I've seen large breaking waves—many of them life-threatening—but never one driven solely by the moon and tide. And never this far from the open sea.

I take one more breath before scrambling up the wall, soaking one of my shoes as I cross the high-water mark. In all my travels, I've never seen a tide so quick, so merciless. On most coasts, it takes six hours for a tide to come in; here it takes six seconds.

By the time the dragon-tide reaches Hangzhou and the Temple of

Six Harmonies, the site of Lu Zheshin's spiritual awakening, it will have traveled about sixty miles in eight hours. The small fishing boats below the temple will nonchalantly turn their bows into the wave, now only inches tall, and let it roll under.

Back in the village of Yanguan, I recognize one of the women who was praying to the river. Like most people I have encountered in the Chinese countryside, she is shy when I approach, but warms up quickly.

"You can go to the pagodas or the Sea God Temple to pray," she says. "But in the end you have to pray to the river. Those other places aren't serious. The river . . . the river is very serious."

4 The Last Magician

Sir Isaac Newton and the
Scientific Revolution

If I have seen a little farther than others it is because
I have stood on the shoulders of giants.

—*Sir Isaac Newton*

The Royal Society of London is one of the only places in the world where you can sign a request slip and fifteen minutes later be handed an original copy of Sir Isaac Newton's *Principia Mathematica*. If you'd like to see something else, like Newton's death mask, you can fill out a slip for that too, but it will take a little longer to deliver.

The society's tall-ceilinged library has rows of bookcases, marble busts of scientists, and, neatly displayed in a glass case, seventeenth-century brass microscopes and telescopes. On the gold-framed coffered ceiling is a painting of angelic women, one lounging with an open copy of Molière on her lap while cherubs flutter nearby. You are not allowed to have pens, briefcases, bags, drinks, or food in here, so when you take your seat, as I do, at the long center table with just a laptop and pencil, there's a feeling of being pared down, unencumbered, ready to begin anew. A sign on the attendant's desk at the head of the room reads "Internet Access Code: Newton + Apple."

There are only three of us in the library. Across the table, a white-bearded gentleman reads a 1670 copy of the *Philosophical Transactions*, a Royal Society publication printed monthly since 1665. A few seats to my right, a young woman studies through a magnifying glass one of James Audubon's original bird sketches. Two coffin-size boxes rest in the corner, cradling pendulum clocks just returned from a museum exhibit across town. Natural light fills the room from three pairs of floor-to-ceiling glass doors, each opening to its own small balcony overlooking St. James Park and the red-graveled mall leading to Buckingham Palace. In the other direction, through leafy and knuckled maples, Trafalgar Square bustles with tourists and local businesses.

Viewed from St. James Park, this building may not be the tallest around, but it is handsome and steadfast. Its three stories of cream-colored stucco are elegantly accented with leaded glass, layered cornices, and crisp wrought-iron railings. At ground level, Corinthian columns align like chess pieces. This address—6–9 Carlton House Terrace—has long attracted international dignitaries. In 1881 it became the Imperial German Embassy, and from 1936 until World War II it was the Third Reich's London headquarters.

The interior is no less regal than the exterior. While waiting for Newton's *Principia* and death mask, I stroll the wide halls and conference rooms adorned with hundreds of oil-on-canvas portraits, many ornately framed in gold leaf and individually lit. In the clerestory above the central marble staircase, a prominent and colorful stained-glass window features the society's coat of arms and motto, *Nullius in verba*. The Latin meaning, "take nobody's word for it," is often distilled to "question everything."

The red-carpeted cafeteria on the bottom floor is rimmed with portraits too. A forty-something Einstein gazes over a buffet of salad greens, fish stew, and fried potatoes. Stephen Hawking, who now holds Newton's position as Lucasian Professor of Mathematics at Cambridge, towers over a display of dessert cakes. Even the less-lighted corners hold surprises. Near the bathrooms, I happen upon a black-and-white photo, dated 1953, of Watson and Crick, smiling nonchalantly beside their spiky head-high model of a DNA double helix.

Originally formed by a small band of scientists, including Christopher Wren, Robert Boyle, and Robert Hooke, the Royal Society of London for Improving Natural Knowledge received its formal charter from King Charles II in 1662. The society's meetings and publications soon became an international forum for scientific inquiry. Weekly meetings and articles in the *Philosophical Transactions* explored topics such as the improvement of optical glasses, a possible cure for the plague, the body's blood circulation, whaling in the Bermudas, and the success (or failure) of pendulum-watches at sea to calculate longitude. Five tide-related articles appeared in the first issue: January 1665.

Newton was invited to attend his first Royal Society meeting in 1668, at age twenty-six. A few years later he became a member, and in 1687 he published the *Principia*. He served as the society's president for twenty-four years, from 1703 until his death in 1727. During Newton's reign, the society continued to grow but struggled to find a permanent meeting place. After his death, the group moved four or five times before signing in 1963 a ninety-nine-year lease at Carlton House Terrace. The address houses the society's 100,000 square feet of offices, meeting rooms, publishing houses, conference halls, and accommodations on the top floor for an occasional overnight visit by one of its 1,500 members.

Opening my computer files charting the society's role in the scientific revolution, and especially Newton's groundbreaking tide theory, I note the conspicuous contributions of a handful of scientists, the "giants" on whose shoulders Newton stood: Nicolaus Copernicus, Johannes Kepler, Galileo Galilei, René Descartes, and a few others, including the society's founders. I'm struck by how little we know of these men outside their scientific contributions. As huge as they were intellectually, many were insecure and jealous, hoarded their findings, and fought bitter personal battles.

The library's silence is interrupted as a heavy oak door opens and someone pushes in a small blue cart with a squeaking wheel from the basement archives. The attendant sets the *Principia* beside my elbow. I run my hand over the 350-year-old tome, wondering what I will learn about Newton and his era when I open it.

Abiding by the rules, I gently place the edition on a bean-filled gray pillow. For a book nearly four centuries old, it's in beautiful shape. The leather is deep maroon and grainy, probably dyed with ox blood. A strip of leather tape connects the spine to the covering boards. The edges are slightly frayed from use, and the top-right corner is worn through, exposing curled inner layers. Its thick pages are spotted, perhaps with coffee, tea, or oily fingers. The page borders are scalloped in places—not torn, but rough-cut and feathered. It smells like an old leather purse or wallet, mixed with hints of acids and glues.

The *Principia* was written in Latin, and publication of its first edition, under the Royal Society's imprint, was financed by Edmund Halley, a Royal Society member and one of Newton's only friends. A few years later, Halley (who, among other things, discovered the comet that bears his name) wrote a brief summary in English for King Henry, which was also published in the *Philosophical Transactions*. A full English translation didn't appear for another forty-two years, in 1729. The scientific world, however, was aching for change and rushed through Newton's door as soon as it was kicked open. For those who could understand (and were looking), Newton's laws of motion and gravity rendered the decisive blow to Aristotle's cosmology. The science of tides, which had been stymied since ancient times by a lack of understanding of how the universe worked, rushed through Newton's door too.

In spite of the *Principia*'s huge import, only three of its thousand pages are dedicated to his tide theory, and even these are largely filled with mathematical equations. Turning to them, I'm struck by the blocky, heavy-lined print against the tobacco-colored paper. There is an elegant flair to the stately letters, curls lifting off the *T*'s and wisps trailing the *S*'s. I don't read Latin, but it makes no difference. Even if Newton's tide theory were expounded in English, I would drown in the equations. I don't feel too bad about this, as Newton's theories were understood by only a handful of the day's most accomplished scientists. Passing Newton on the street, a Cambridge student was said to have whispered, "There goes the man that writt a book that neither he nor anybody else understands!" The people who were watching and listening, however, understood that between *Principia*'s covers was a neatly laid-out description that would forever and decisively change the world.

Newton's tide theory squeaked into the seventeenth century as haltingly and unceremoniously as the blue cart that has just delivered his book from the archives. There were huge gaps to bridge between antiquity and the modern Newtonian world—and not just in terms of scientific discovery. First, the Aristotelian-Ptolemaic "system of the world" had to be dismantled. As long as the earth and not the sun was perceived as the stationary center of the universe, the tide's cause could never be found. Second, once the sun was put in its rightful place, the laws of planetary motion had to be laid out. And third, if the moon was queen of tides, how did she do it? By what mechanism? Did she throw or push something against the oceans, like someone might throw a ball at a window or push against a door? Or did she accomplish her work with no visible mechanism, as if reaching through space with ghostly fingers? If the latter, how could a mysterious and unseen force be reconciled with the emerging Newtonian clockwork universe?

⌁

For thousands of years, nearly everyone saw the universe through Aristotle's eyes. His system of perfect spheres, fortified by Ptolemy's earth-centered universe, was the foundation on which Western culture and theology was built. In Aristotle's view, the universe was made of concentric circles emanating from inside the earth and extending far into the heavens. The circles, or spheres, between the earth and moon were considered impure and changeable; everything beyond the moon was considered perfect and unchanging. The farther into the heavens, the more perfect and unmoving the spheres. Hence, the Primary Mover—the source of all movement—was located in the farthermost sphere. For Christians, this was the ideal place to put God, and the

*SCHEMA PRÆMISSÆ
DIVISIONIS.*

B

Aristotle's spherical universe, depicted in this 1524 sketch, dominated the Western worldview until the end of the seventeenth century. Note the earth at center and the "primary mover," or God, on the outermost sphere.

ideal place to put mankind was on earth, the center of the universe. The earth, however, was also one of the lowest and most corrupt spheres, which fittingly described mankind's plight of original sin, redeemed only through atonement with the upper—heavenly—spheres.

In the early sixteenth century, Copernicus proposed that the sun, not the earth, was the center of the universe and that the earth moved around the sun just as the other planets did. Although sun-centered theories were not new, Copernicus was the first to work out the math and make it public. Knowing how unpopular his proposal would be, he wrote in the preface of his *Book of Revolutions* (which was dedicated to Pope Paul III): "I may well presume, most Holy Father, that certain people, as soon as they hear that . . . I ascribe movement to the earthly globe, will cry out that . . . I should at once be hissed off stage." Indeed, the *Book of Revolutions*, which was published in 1543, the year Copernicus died, *was* hissed offstage as soon as the Church understood its implications. But the damage was done. The scientific revolution was underway. Kepler and Galileo followed shortly, each delivering another blow to the ancient system and each suffering the Church's wrath in consequence.

A generation after Copernicus, the Danish astronomer Tycho Brahe (1546–1601) built an observatory on the island of Hven and began collecting data on the movement of stars and planets. He was not a Copernican, but his data—the most thorough and trustworthy of the era—served those who were, especially a young German astronomer and mystic named Johannes Kepler, who assisted Brahe for about two years. They didn't get along, but when Brahe died, Kepler gained access to his voluminous—and until then private—records, using them as a stepping-stone to develop his three groundbreaking laws of planetary motion.

Born in 1571 near the Black Forest, Kepler was a sickly child with spindly limbs and a pasty, disproportionately large face. A bout with smallpox at four left him with crippled hands and impaired vision. He could focus close up but not at a distance, often seeing the world dou-

bled or quadrupled. At thirteen he began theological studies. Ten years
later, discovering an aptitude for science, he accepted a position teaching mathematics and astronomy in the Austrian city of Graz. Although
he chose a career in science, mysticism and spirituality remained central interests. He sought the universe's mathematical plan, believing
that every equation, every theorem, was God's creation.

Astrology, too, was mixed in. Kepler wrote horoscopes and published astrological calendars to supplement his income and spent
his last years as a court astrologer in Prague. He called astrology a
"dreadful superstition" but also recognized it as a means of relating
the individual to the universal whole. "The natural soul of man," he
wrote, "is not larger in size than a single point, and on this point the
form and character of the entire sky is potentially engraved."

Kepler's first book, *The Sacred Mystery of the Cosmos*, was published
in 1596, when he was twenty-five. In it—against the advice of his colleagues—he supported the Copernican system outright, describing it
as "an inexhaustible treasure of truly divine insight." Copernicus had
died fifty years earlier, yet no professional astronomer until Kepler
had dared to publicly support his theory. Even Galileo, Kepler's senior
by six years, refused to come out of his Copernican closet until he was
almost fifty. Who could blame him? About this same time the Holy
Office of the Inquisition charged Giordano Bruno, a Copernican and
pantheist, with multiple counts of blasphemy and burned him alive
in Rome's Square of Flowers. Bruno was one of the first scientists to
suffer this fate. His books were immediately put on the *Index Librorum
Prohibitorum* (List of Prohibited Books), where they shared at one time
or another the good company of works by Copernicus, Kepler, Galileo,
Descartes, Hugo, Bacon, Milton, and scores of others. The Church's
message resounded; its insecurity haunted the era's great iconoclastic
thinkers. Kepler was excommunicated in 1613, and Galileo was sentenced to house arrest in 1633. Both Descartes and Newton foundered
under the weight of the Church.

Kepler's attraction to heliocentrism (a sun-centered universe) was more a matter of metaphysics than astronomy. He saw the universe as the Holy Trinity's symbol and signature—the sun representing the Father, the sphere of fixed stars representing the Son, and the invisible forces (emanating from the sun) representing the Holy Ghost. In his vision, there was no question that the sun—the Father—belonged at the universe's center.

Textbooks tell us that Kepler's three laws of planetary motion were pillars in Newton's theory of gravity. They don't commonly mention that Kepler himself came close to discovering gravity—almost a hundred years before Newton. As an extension of his first law, which described planetary orbits as elliptical rather than circular, Kepler noticed that planets orbited the sun at different speeds. The closer a planet was to the sun, the faster it orbited. The reason, he suspected, was that the sun exuded some kind of force felt more strongly close up, just as a light appears brighter the nearer we are to its source. Planets farther away, such as Saturn, move more slowly because they feel the force more weakly.

If the sun emits an attractive force, why wouldn't the planets fall into it? Kepler figured there must be a resistive force within each planet, which he called "laziness." He saw this explicitly at work in the earth's oceans, which would logically spill into the heavens if there weren't a certain laziness keeping them pinned to the planet. The combined push and pull of these forces, he surmised, caused the tides.

The idea of a "force" was not new. Naturally occurring magnetic rocks, or lodestones, were well known to the ancients. The word "magnet" comes from *lithos magnetis,* or "magnesian stone," which may refer to lodestones found near Magnesia, Greece. In ancient times, the lodestone was a source of superstition and exaggeration. Pliny the Elder wrote: "Near the river Indus there are two mountains, one of which attracts iron while the other repels it. A man with iron nails on his shoes cannot raise his feet from the one or put them down on the other."

The first reference to the use of a magnetic compass for navigation appeared in the eleventh century. This invaluable tool was most likely discovered in China and later, independently, in Europe. Sailors grew familiar with the miraculous north-pointing needle, but debates ensued as to whether the needle was *pulled* northward (by the pole star or a large magnetic island) or simply *pointed* in that direction.

The British astronomer William Gilbert published *De Magnete* in 1600, in which he described the earth itself as a large magnet, possibly with a "field" or "virtue" stretching beyond the atmosphere. Kepler, who was familiar with Gilbert's work, figured this field was reciprocal: the earth attracted the moon, and the moon attracted the earth. He also speculated that the sun's attractive force might be felt all the way to the earth, and hence both the sun and moon played a role in the ocean's tide.

These "fields" or "forces" were, in Kepler's initial view, the invisible workings of the Holy Ghost. Later, he called them "soul" and sometimes *vis motrix* (life force). He was one of the few early philosophers who was comfortable with the idea of "action at a distance"—that an object could influence another across space or time without an apparent physical connection or mechanism. No matter the name—and no matter that he failed to develop the idea further—the concept of action at a distance was a dramatic departure from the old paradigm. In Aristotle's cosmology, heavy bodies moved downward by their *inner nature*. A stone thrown in the air falls back to earth because *that's its nature*. Aristotle called this *gravitas*, or heaviness (the word *gravity*, until Newton got hold of it, meant "weight"). Elements such as fire and air, being light, moved upward because of their *inner nature*. There was no such thing in ancient cosmology as an external "force" or "emanating energy." Kepler's concept of forces, or "soul," was radically new.

Also new, and subtly prescient in our tide story, is Kepler's discovery of the solar system's musical proportions. In *The Harmony of the World*, published in 1619, he suggests a harmonic ratio between a planet's slowest and fastest orbiting speeds. Saturn's ratio, for instance, is 4/5, almost a perfect *minor third*. These are real harmonies but soundless, at least to us. Kepler even conjectured that, once in a while, the harmony of two planets would converge, and once in a greater while, the harmony of *all* the planets would converge. "If there could occur one six-fold harmony," he wrote, referring to the six known planets, "that undoubtedly could be taken as characterizing Creation."

Much later, scientists would discover that harmonics—musical vibrations—are the tide's essence. The discovery of gravity had to come first, but eventually the repeating orbiting patterns of the sun and moon would be understood as a kind of orchestra, played to an intently listening sea. If Kepler's six-fold harmony failed to ignite the fireworks of Creation, it would at least excite an exceptionally high tide.

⁓

I wish I could meet Kepler, but the best I can do is stroll the Royal Society halls in search of his portrait. I pass a clean-shaven and youthful Copernicus and an officious-looking Brahe adorned in a long handlebar mustache and spectacles. His nose, which was sliced off in a sword fight, is capped with shiny steel. A dark-cloaked Sir Francis Bacon looms nearby. One could get lost in these halls—not physically, but temporally. Captured by the intent gaze of Christopher Wren, I feel a chink loosen in what separates us. In this building, differences in time and distance dissolve. For now, their world is my world.

I find no Kepler. Instead, I come upon his Italian colleague, Galileo Galilei. The painting, a two-by-three-foot, oil-on-canvas bust, is a near-perfect copy of an original by Justus Sustermans that hangs in the Uffizi Gallery in Florence. Galileo had commissioned the painting in 1636 as a gift for a Parisian friend. He was seventy-two at the time, under house arrest at his Arcetri farm. His bearded face, life-size and flush with color, turns to the right and upward. His wistful green eyes gaze in that direction too. Moisture pools in the right eye, which went blind from a combination of cataracts and glaucoma soon after the portrait was finished. He was completely blind by year's end, never seeing or reading the published version of his last book, *Two New Sciences*.

Unlike many nearby portraits, Galileo is not holding a book, an instrument, a document, or something else that indicates his social status or occupation. Nor does he appear stern, pedantic, or distant; instead, he is gentle, pensive, even surrendering. During those years, he confided to a friend, "I feel immense sadness and melancholy, together with extreme inappetite; I am hateful to myself." And to his beloved daughter, Sister Maria Celeste, he wrote that he felt as if his name had been stricken from the roll call of the living.

Dialogue on the Great World Systems, Galileo's third book and the one that got him into trouble with the Church, had been released a few years earlier, in 1632. The book tracks a theatrical encounter among fictional characters over four days. The first two days are a debate of Aristotle's system. The third day lays out the arguments for and against Copernicanism. On the fourth day, Galileo rolls out his tide theory, claiming that it provides final proof that the earth rotates on its axis once a day and revolves around the sun once a year. Galileo evidently developed his tide theory while sailing on a boat hauling fresh water to Venice. He noticed that the water in the jugs sloshed with the boat's jerking movement. From this, he extrapolated that the oceans, like the water in the jugs, must slosh to and fro with the unsteady rotation of the earth. It was a brilliant insight. Not only did it provide a tangible tide-generating mechanism but also elegant proof that the earth *moved*. Since antiquity, astrologers had pointed to the tides as proof of divine influence on earth; now Galileo pointed to them as proof of a new cosmic order.

Galileo's tide theory was so central to his heliocentric argument that his book's original title was *The Flux and Reflux of the Tides*. Pope Urban VIII thought the title gave too strong an impression of "physical proof" (rather than mere "hypothesis") of heliocentrism and recommended it be changed to *Dialogue on the Great World Systems*.

The book sold out almost immediately in Florence. Although an outbreak of the plague delayed its release in Rome, it sold well there too. But it was dripping with heresy. Galileo had been warned sixteen years earlier, in 1616, "not to hold, defend, or teach" the Copernican view of a sun-centered universe. *Dialogue on the Great World Systems* flew in the face of that edict. Within a few months of the book's release, the Catholic Church issued an official order banning the book and demanding that the author appear before the Holy Office of the Inquisition. Due to poor health, Galileo didn't make it to Rome until October. Eight months later, on June 22, 1634, the cardinal inquisitors gathered in the Dominican Convent of Minerva in the city's center and publicly convicted Galileo of heinous crimes. Draped in a white robe, Galileo kneeled on the cold stone floor and abjured. He was sentenced to house arrest for the rest of his days, forbidden to see anyone with whom he might discuss science.

Of this tragedy, the historian Thomas Kuhn wrote, "No episode in Catholic literature has so often or so appropriately been cited against the Church as the pathetic recantation forced upon the aged Galileo." Three hundred and sixty years later, in 1993, the Catholic Church formally acquitted him.

~

Galileo appears to have developed his tide theory only to the extent necessary to make a case for heliocentrism. He hardly mentioned the widely acknowledged role of the moon. Galileo's most important scientific contributions were in other fields, such as the dynamics of inertia and acceleration and his discoveries while peering through a telescope. Nevertheless, his tide theory had two essential components. First, when he observed the water sloshing in the shipboard jugs, he suspected that each jug responded differently to the boat's motion, depending on its size, shape, and how much water it contained. A small jostle might excite a lot of motion in one jug but not much in another.

Seeing this, Galileo reasoned that each jug had its own "natural period of oscillation" or pattern of vibration. If this were true, he reasoned, then each ocean, like each jug, would respond differently to the earth's movements.

The idea was at least a hundred years ahead of its time. We now know that ocean basins do indeed have unique vibratory patterns, defined by their shape and depth and even their saltiness. The Pacific, by far the largest ocean, responds differently to the sun and moon than, say, the Atlantic or the Mediterranean. Each ocean sloshes to its own rhythm, just as the water jugs did.

The other noteworthy element in Galileo's theory was his insistence on identifying a visible and tangible mechanism—the earth's motion—as the tide's cause. There was no room in his view for unseen or mysterious forces. As in a game of billiards, balls move because—and *only* because—they are struck by the cue ball. It makes eminent sense. In Kepler's vision, there was no cue ball. At least not a visible one.

Galileo was appalled by Kepler's notion of a "soul" or "life force" reaching out unseen over vast distances. "That concept," wrote Galileo in *Dialogue*, "is completely repugnant to my mind." And, a few pages later: "I am more astonished at Kepler than at any other. . . . Though he has at his fingertips the motions attributed to the Earth, he has nevertheless lent his ear and his assent to the moon's dominion over the waters, to occult properties, and to such puerilities."

Galileo wasn't the only prominent thinker appalled by notions of an invisible force. His French counterpart, René Descartes (1596–1650), offered a tide theory reminiscent of the ancient idea of a "whirling vortex." He envisioned Aristotle's fifth dimension, made of ether, as a thick, pervasive soup that "compressed" against the earth and its oceans as the moon circled overhead. In his mind, the ether was a tangible mechanism capable of influencing objects over vast distances. Peculiarly, this would mean the tide was "compressed"—or low—when the moon was overhead. It made no practical sense, but

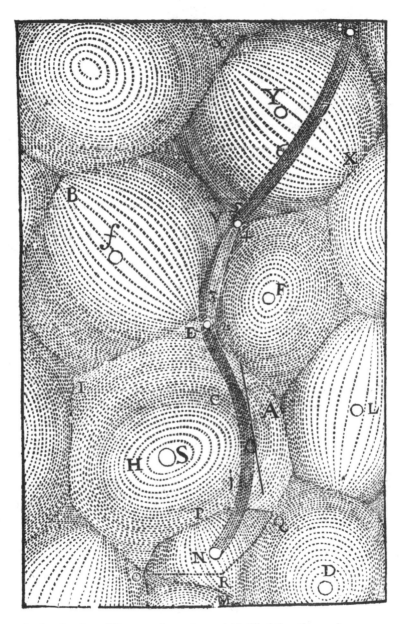

A 1692 drawing of Descartes's vortices, which filled the ether and explained how the moon communicated its force—from one vortex to the next—to influence the ocean's tides.

Descartes's theory remained popular—especially in France—decades after Newton's theory of gravity had gained acceptance throughout most of Europe.

On the way back to the library, I'm startled by Descartes's caricature-like portrait. There is nothing quite so odd hanging on the society's walls. The painting, a copy of an original at the Louvre, is dark except for the face, collar, and a few pasty fingers curling around the rim of a felt bowler at his waist. His black hair is shoulder-length, with bangs framing his face. Somber brown eyes stare from under drooping lids. He is clean-shaven except for a wispy mustache and a patch of hair hanging off his lower lip. His thick eyebrows arch as if he has asked a question and dares a reply. I study the image for a few minutes, wondering what the question might have been. History knows him as the father of mechanistic philosophy, which posited, among other things, the separation of mind and body. Descartes was also an accomplished mathematician, poet, and mystic. It's hard to believe, with such a pedigree, that he would come to doubt the existence of everything but *thought*. In essence, "I think, therefore I am" is a denial of all sensory experience.

Perhaps the significance of Descartes's thinking can be seen more as a rebellion against classical views than something of enduring insight. Like Galileo, Kepler, and Newton, he lived precariously between the ancient and modern worlds, not fully in one or the other.

The science of tides languished between these worlds too. Although nonlunar tide theories surfaced here and there, most natural philosophers by the mid-seventeenth century believed the moon was the tide's primary cause. Yet no one knew how it did it. *Yes*, the tide rose and fell as the moon traversed the sky; *yes*, there were spring tides at new and full moon; *yes*, there were neap tides at quarter moons; and *yes*, there were longer lunar cycles, such as perigee and apogee, that influenced the tide. Still, how did the moon do it? Where was the cue ball? As late as 1650, the German geographer Bernhardus Varenius wrote:

Some have thought the Earth and Sea to be a living creature, which by its respiration, causeth this ebbing and flowing. Others imagine that it proceeds, and is provoked, from a great whirlpool near Norway, which, for six hours, absorbs the water, and afterwards, disgorges it in the same space of time. . . . Others supposed that it is caused by the opposite shores, especially of America, whereby the sea is obstructed and reverberated. But most philosophers, who have observed the harmony that these tides have with the Moon, have given their opinion that they are entirely owing to the influence of that luminary. But the question is, what is this influence? To which they only answer, that it is an occult quality, or sympathy, whereby the Moon attracts all moist bodies. But these are only words, and they signify no more than that the moon does it by some means or other, but they do not know how: Which is the thing we want.

"The thing we want" turned out to be gravity, but we had to wait for Newton to get it. Just like Copernicus's heliocentrism, which was initially hissed offstage, it took a long time for gravity to gain acceptance. The idea of a ghostly force spreading its fingers across vast distances was so unscientific that most natural philosophers wanted nothing to do with it. In a world yearning for mechanistic answers, gravity seemed a regression into the occult of antiquity. Even Newton disliked it. He wrote in a famous letter to Richard Bentley, one of England's leading theologians and master of Trinity College: "Tis inconceivable that inanimate brute matter should . . . operate upon and affect other matter without mutual contact. . . . And this is one reason why I desired you would not ascribe innate gravity to me. That gravity should be innate . . . is to me so great an absurdity, that I believe no man who . . . has a competent faculty of thinking can ever fall into it."

Newton tried to rid himself of this strange force. Somewhat embarrassed, he ultimately found no other solution than to plunk it into the center of his universal theory. He never discovered what gravity was, only how it behaved.

We still don't fully know what it is.

≈

Robert Hooke, an extraordinary scientist and secretary to the Royal Society (and, later, president), gave a lecture to society members in 1666 in which he used the word "gravity" as we presently understand it. In his "System of the World," Hooke claimed that all bodies have "an attraction or gravitating power towards their own center." Hooke even proposed that gravity is proportional to an object's mass and inversely proportional to the distance between masses. The subject failed to ignite much interest, either because it was too far-fetched, too hard to understand, or too distasteful. But Hooke was right.

At this same time, Newton was beginning to suspect the presence of a ubiquitous force like gravity too. In the spring of 1665, as the plague crept across Britain, snuffing out hundreds of thousands of lives by year's end, Newton fled Cambridge for his home in Lincolnshire. He was there only eighteen months, but it was one of his most productive periods. He made momentous forays into optics, mathematics, and chemistry and invented the calculus that he would later use to support his theories of motion and gravity. Ever the secretive soul, Newton kept these discoveries to himself, only coming forward years later at the coaxing of friends, including Edmund Halley, the philosopher John Locke, and the diarist Samuel Pepys.

It was Halley who finally convinced Newton, in 1684, to finish and publish his work on gravity. What Newton had that Hooke lacked was a mathematical foundation. He was able to give teeth to an otherwise gummy notion. Newton saw the ocean's tide as a critical puzzle piece that would have to be reconciled with all other cosmic motion.

Hooke was indignant when Newton's paper was first read to the Royal Society. Ten years earlier, the two men had fought bitterly over contrary theories of light, and now they would battle publicly for ownership of the universal laws of motion and gravity. In fact, due to this antagonism, Newton stayed away from the Royal Society until Hooke's death in 1703, at which time Newton assumed the presidency. One of his first official gestures was to destroy Hooke's portrait. Intellectual giants, again and again these men proved to be moral and emotional dwarfs.

At my library seat, I continue reviewing notes and drawings collected over several years of research. Thankfully, most are written in English. The elegance of Newton's equilibrium tide theory is that it strips away the layers to reveal the simplest picture, then slowly puts them back in. The theory begins with a hypothetical water-covered planet with no continents. This immediately dispenses with the messy interaction of ocean waves as they bounce off landmasses, double back on themselves, slow down and steepen in shallow water, and speed up and flatten in deep water. By proposing a uniformly deep ocean, Newton also eliminates the tide's most frustratingly complex element: shallow water. In shallow water, all logic about tide behavior goes awry.

To simplify further, he banishes the sun, leaving only the earth and moon. With just these two bodies, Newton was able to demonstrate how gravity causes the tides, a theory that in essence holds true today.

The movements of the earth and moon—along with their eccentric yaws, wobbles, precessions, and varying speeds—are too complex for an ordinary person to hold in mind all at once. Most of us can readily envision the earth spinning every twenty-four hours (the solar day) or revolving around the sun every 365 days. We can also envision the moon orbiting the earth about every twenty-nine days—the lunar month. But we often forget that the earth and moon also circle *each other*. Locked together by gravity, they spin around a common center, neither flying away nor falling into one another. Gravity keeps them from flying away, and an equal and opposite force keeps them from falling into one another. That opposite force—what Kepler called "laziness"—is not lazy at all. Newton called it centrifugal force, which pulls a rotating body away from the center of rotation (from Latin, *centrifugal* means "to flee from center"). We experience this force when driving a car around a curve, especially if we don't slow down. We hold

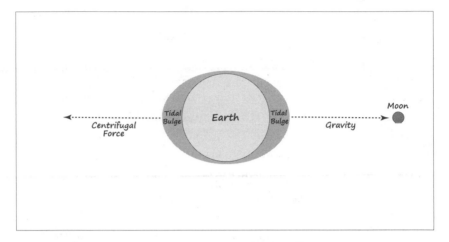

Two tidal bulges, one drawn by the moon's gravity and the other "fleeing" due to centrifugal force. They are always equal and opposite.

the steering wheel (as in "holding the turn"), but the car wants to flee from center, straighten out, or, worse, roll over.

Gravity and centrifugal force *exactly* counteract each other, keeping the moon and earth in a stabilized orbit as they spin together through space. But under the influence of these two forces, the ocean, being more fluid than the solid earth, responds differently. The waters directly under the moon are drawn up by gravity, while on the opposite side of the earth, the ocean wants to flee, so it bulges outward. It responds like a car that can't hold a turn. The result is two bulges of equal size, one on the side of the earth facing the moon and one on the opposite side of the earth (see figure above). Each of these bulges is what we experience as high tide.

Another way of looking at this is to imagine swinging a tethered water balloon in a circle. As you swing, the balloon stretches, creating two bulges, one pulled outward by centrifugal force and one pulled inward by the tether (think of the tethered end as gravity). Under the constant influence of these two forces, the earth's oceans look just like

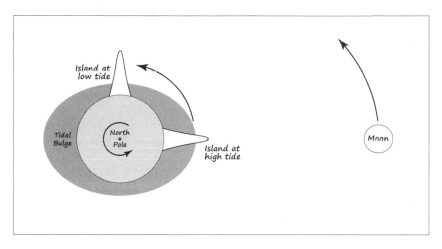

Looking down on the earth, an island "under" the moon would experience high tide. As the earth rotates, the island would see a low tide about six hours later. About six hours after that, it would see another high tide, and so on.

a stretched out water balloon. The stretching out represents a state of equilibrium between two opposing forces, hence the theory's name.

To understand why there are generally two high tides and two low tides a day, let's place an island in this hypothetical landless ocean—right under the moon (see figure above). At that location, the island experiences high tide. As the earth makes its daily rotation, the island naturally rotates with it from one bulge to another and through the "valleys" in between. The valleys are low tides. In increments of roughly six hours, the little island will first experience high tide, then low, then another high and another low as the earth rotates under the tidal bulges. This pattern—semidiurnal—occurs on most of the world's coasts.

If the moon can cause such a stir in the ocean, wouldn't it also affect the solid earth? Do our bodies, made of 70 percent water, have a tide?

Yes and no. The earth is as rigid as a steel ball, but it does distort under the gravitational influence of the sun and moon. High tide on the solid earth varies from half a foot to three feet and spreads over such a large distance—about ten thousand miles—that it's not perceptible. For example, a high spring tide might raise the sidewalks and buildings of New York by a couple of inches. You could never detect this as you walk down Broadway, because *everything* rises and falls together over a six-hour period (unlike on the coast, where the ocean rises and falls relative to the beach). The tide's daily squeezing and releasing of the earth has long been known to affect water wells too. A Wisconsin well, about eight hundred miles from the nearest ocean, has a two-inch tide. An inland well in France increases its flow from sixty to ninety gallons an hour during spring tides.

We have a tide in our bodies too (and even in our morning cup of coffee), but it's barely discernible. In fact, our body weight changes by less than a drop of sweat when the moon is overhead. This is because our body mass is small and the tide-generating force is actually quite weak. Pervasive, but weak. The moon and sun's gravitational forces have a hard time lifting things straight up, a much easier time pulling things across the surface. To get a feel for this, imagine the difficulty of lifting a rowboat vertically off the water as opposed to pulling or pushing it across the surface.

In a sense, the moon isn't strong enough to lift the oceans directly upward. Instead, it pulls water across the earth's surface, as one would rake leaves. Eventually, the leaves are raked into a pile directly under the moon. This pile is one of the bulges we experience as high tide.

While gravity is raking water toward the moon, centrifugal force is raking it away. Each force keeps its own pile as the system whirls through space.

Before putting the sun back into Newton's model, I take a break and go looking for his portrait. There are several hanging on these walls, more than forty worldwide. The one I come upon is dated 1627, the year he died. He is sitting in a tall-backed green chair set at an angle to the left while his face is turned right. A dark overgarment loosely covers his upper body, with a white silk shirt exposed at the chest and wrapped high about the neck. The light comes from the right of the image, drawing our attention to his face and delicate hands as they emerge from the flared white cuffs of his cloak. The hands are pale and blue-veined, with long, slender fingers. The right hand rests on the chair's wooden arm, outstretched except for a slight bend in the index finger. The left hand drapes over a book's spine, presumably the *Principia*. I imagine we are meant to notice that these are not the hands of a laborer.

Newton was born in a small stone farmhouse on Christmas Day 1642—the year Galileo died. He was so small and frail that nursemaids who were sent to get help dawdled because they saw no point. His father, Isaac, had fallen ill and died before his son's birth. When Newton was three, his mother, Hannah Ayscough, married a man who didn't want a stepson, so she left young Newton with his grandmother. He was expected to grow up tending the family farm. When Newton showed no interest or talent for it, an uncle sent him off to school. At nineteen, with the help of another relative, Newton enrolled at Cambridge University's Trinity College. Cambridge students were recognized in three categories: the elite noblemen who received degrees with little examination; pensioners, who paid room and board and aimed mainly for a ministry degree; and sizars, the lowest class. Although Newton's mother was wealthy by then, she gave him so little financial support that he began his student life as a subsizar. As such, he served other students, ran errands for them, ate their leftovers, and emptied their chamber pots. He felt lucky, however, to have a desk, a notebook, some ink, and candles to light the long nights.

With the exception of a few trips back to his hometown—one to escape the plague and one to care for his dying mother—he spent much of his life at Cambridge. There his genius surfaced in everything he turned his attention to: mathematics, light and color, physics, theology, alchemy. At the age of fifty-four, he moved from Cambridge to London to supervise the manufacturing of coins for currency, a position appointed by the king.

For a man who lived in a nation surrounded by water and who

authored a revolutionary tide theory, it's curious that his path across the earth's surface never ventured farther than the 150 miles of rutted roads between Lincolnshire, Cambridge, and London. Most likely he never saw the ocean.

In the portrait, Newton's cheeks and chin are jowly. As Master of the Mint, he was wealthy by this time. A brown-blond wig, parted in the center, wraps his face and unfolds its curls on either side of his chest. His large hazel eyes gaze off. At first he appears distant and cold—not of this world. But the longer I study, the more wistful he seems. The portrait's background is dark and featureless except for the artist's signature—Van Der Banck—and, off Newton's left shoulder, a snake biting its tail, the symbol of eternity.

⁓

Newton left the sun out of his initial model for good reason. The sun is so big (416 times larger than the moon), it's hard to imagine Newton's perfectly balanced model containing it. Yet by the natural philosopher's own second law of motion, gravity varies in strength according to the mass and distance between objects. The sun's overwhelming mass isn't enough to overcome the fact that it's 93 million miles away. The moon is only 239,000 miles from earth. This difference in distance is the reason the sun's tidal influence is only half that of the moon's. This is still significant, so when the sun is put back into Newton's model, it *does* change things.

Like the moon, the sun creates two bulges in the earth's oceans. Because the oceans are fluid, the sun's influence either adds to or subtracts from the moon's bulges. When the sun, moon, and earth are aligned (in syzygy) the moon and sun are working together, creating a superbulge. That's when we experience spring tides.

When the moon is half full, either waxing or waning, it's 90 degrees from the sun, and the two are working at odds (see figure on page 36). The sun steals some of the moon's pile of leaves and vice versa, leaving them both with smaller piles (or a displaced pile somewhere in between). This is when we experience neap tides, about every fourteen days.

Newton's model showed how the moon and sun created two high tides and two low tides daily (semidiurnal) of roughly equal size, which generally matched observations in the Atlantic. But by Newton's day, sailors were embarking on worldwide expeditions and bringing back tales of unusual tides. On some shores, for example, the tides were uneven in height. On any given day, one high tide was higher than the next and one low tide was lower. This phenomenon—called *semidiurnal inequality*—was witnessed in ancient times, but now it was confirmed. World sailors saw it all along the west coast of North America, from California to Alaska. But why?

Newton provided the answer in the last brushstroke of his equilibrium theory. In his simplified model, the moon is aligned with the earth's equatorial plane. But just as the earth is tilted 23.5 degrees from that plane relative to the sun (the reason we have seasons), so the moon is "tilted" in relation to the earth. What this means is that the moon isn't usually aligned with the equator but moves north or south of it. *Moves* is an understatement. In a month's time it darts like a hummingbird from 28 degrees north of the equator to 28 degrees south of the equator and back again. That's 112 degrees of change per month. The earth logs two degrees during the same period (relative to the sun). We can get a glimpse of the moon's hummingbird-like behavior by watching how it rises each night at a different location on the horizon.

Newton figured that the moon's migration north and south of the equator (its changes in *declination*) must play a role in the varying heights of daily tides. In his first stripped-down model, the moon was frozen on the equator. Next he froze it north of the equator (see figure on next page). When it's there, the ocean bulges are aligned such that one is in the northern hemisphere and one in the southern hemisphere. These bulges are not aligned with the rotation of the earth, so a fixed object doesn't pass symmetrically from one bulge to the next,

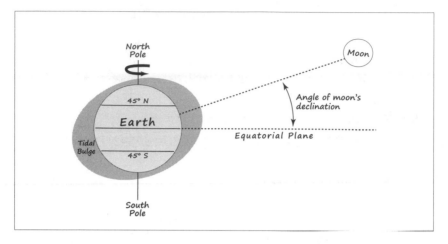

When the moon is north or south of the equator, the tides are often unequal, meaning the day's high tides and the day's low tides are of different heights (called *semidiurnal inequality*). In the figure above, imagine standing at 45 degrees north latitude and circling from the large bulge under the moon (highest high tide) through the edge of the opposite bulge (a lower high tide). The tide's inequality is also affected by the sun's position north or south of the equator.

but instead passes through only one bulge fully while catching just a corner of the next. This is why the two daily tides are often uneven in height, why a day will usually bring a sequence like this: high high tide, low low tide, not-so-high high, and not-so-low low. This inequality will fluctuate as the moon migrates, being most extreme when the moon is farthest north or south of the equator, and least when the moon is on the equator. In places that experience extreme inequality, such as Los Angeles, the not-so-high high and the not-so-low low can be almost identical in height. This is called a *standing tide*, because the water appears to stand still for twelve hours.

When the sun returns once again to the picture, you can imagine how its apparent migration north and south of the equator has a similar "unequalizing" effect. When both the sun and moon are north or south of the equator, as they often are during summer and winter solstice, diurnal inequality is at its greatest.

While mentally grappling with these gyrations, I didn't notice that a gray cardboard box had been delivered to my table. If I weren't in the Royal Society's library, I would take it for a shoebox of memorabilia found in my grandmother's attic. I almost push it aside but catch the penciled letters: Newton's Death Mask.

Really? He's *inside* there? I pull the box over and scan the room, sure that someone will stop me. I'm tempted to spirit the box into a private room where I can be alone with it, where I can savor every second, every touch, every detail without interruption. I can't, of course, leave the room with anything but my computer and pencil, so I calm down and try to act as if the whole experience is routine.

At eight inches wide, eleven inches long, and four inches deep, the box can't weigh more than five pounds and is nondescript except for lettering on the sides and top—all in caps, all the same three words. One side is marked in ink, neater, and another side in black marker, a little looser. The box edges are worn and peeling.

Inside, the mask is wrapped in heavy gauze. I peel back a thin panel and expose the mottled bronze-colored face. His eyes and mouth are closed, tenseless. Other than a broken edge on the left side and a couple of flat spots—one dime-size on the tip of the nose and one quarter-size above the left eyebrow—the mask is in good shape. Some say the face is scowling, but I don't see that. The impression may come from the sunken eye sockets, strong nose, and bony furrows above the eyebrows—all probably accentuated by loss of weight during the final months of life. Other features—the broad chin and gentle curves of the mouth and eyes—convey a softness, a tranquility.

No one knows for certain who prepared the mask, but it was done when Newton's body was still warm, before he was carried to his tomb at Westminster Abbey. The process began by ladling soupy plaster over his face. A centerline thread was embedded from forehead to chin before applying several more layers of thickened plaster. The thread

Newton's death mask.

was pulled before this hardened, dividing the mask in two. The halves were left on the face until the plaster set up, then the two halves were removed and fitted together. The inside was cleaned, smoothed, and filled with plaster to create the final product. In Newton's case, the mask was used as a template for sculpting his tomb and carving a marble bust for Trinity College.

The mask eventually fell into the hands of eighteenth-century French sculptor Louis-François Roubiliac, who carved a statue in its likeness that now stands in Trinity's College Chapel. After being sold at auction in 1762, the mask sat unnoticed for almost eighty years in a dealer's shop. A Royal Society member, Samuel Hunter Christie, discovered it in 1839 and donated it to the society.

I'm not allowed to touch Newton's face, but having spent hours with his theories and books, I don't feel the need. In some way, I am close enough—close enough to glimpse the man, cracked and imperfect like his mask, yet full of extraordinary humanity. Like other intellectual giants, Newton was unimaginably complex, and what history has made of him is even more complex. What are we to believe? What are we to feel toward the man himself? Even his ideas have been culled and trimmed and interpreted, until what any schoolchild knows is simply that he discovered calculus, the laws of motion, and gravity—by an apple falling on his head. Yet in these halls and in my research, he comes alive with mesmerizing capacity, contradiction, loneliness, yearning. He was shy, secretive, afraid of confrontations; he had few friends and never married. If recent evidence proves true that he was a homosexual, that would have added another layer of secrecy, as homosexuality in his day was a crime punishable by death. Newton also fought vitriolic battles, not just with Robert Hooke but with John Flamsteed, the first appointed Astronomer Royal, and the German philosopher Gottfried Leibniz (over the invention of calculus). In later years, as Master of the Mint, he pursued counterfeiters with grisly ruthlessness, sentencing to death one man he had hunted for years.

A pious man, Newton studied theology as intensely as he did physics, yet as a member of a secret society of alchemists, he also cooked potions over a coal fire in search of the elixir of life that would convey immortality.

The poet William Blake (1757–1827) called Newton "a cold rationalist." But like most geniuses, including Blake himself, Newton was too enigmatic to describe definitively as *this* or *that*. He unlocked the door to the modern world, yet his delvings in theology and alchemy, and even his bold acceptance of the invisible and mysterious force of gravity, were rooted in antiquity. Newton walked in both worlds—the old and the new. I like what his biographer, John Maynard Keynes, said of him: "Newton was not the first of the age of reason, he was the last of the magicians."

5 Big Waves

Surfing Mavericks and Nineteenth-Century Tide Theories

Should we slink back inside to our reliable equations
and brood over the inconsistencies of nature? Never!
Instead we must become outdoor wave researchers.
It means being wet, salty, cold—and confused.

—*Willard Bascom*

The surf break at Mavericks, just south of San Francisco, springs to life only a few times a year during the largest Pacific swells. In fact, Mavericks, a tabletop reef that looms a half-mile off Pillar Point near the town of Half Moon Bay, is usually so calm that local fishing fleets sail over it without notice or bother. But when swells twenty feet or larger roll in from distant storms, the place transforms into a tempest. These conditions drive away fishing boats but beckon surfers from around the world to compete at Mavericks's yearly big-wave contest.

The contest, slated each year between November and March, occurs whenever a North Pacific storm stirs waves large enough. The bigger the waves, the better. Event organizers spend their winters studying storm and wave models. If they see the right statistical lineup—a blend of tide, wind, and swell direction—they announce the contest, leaving only a day or two for contestants to fly in from around the world.

Storms and waves are fickle, however; there's no guarantee that any given winter will deliver the right surf conditions. The winters of 2011 and 2012 never saw waves large enough to hold the contest. As the 2013 season opened, contestants and organizers hoped for a change.

On January 18, 2013, I received an email alert that a ferocious storm was kicking up fifty-foot waves in the Bering Sea. The storm was 1,500 miles away, but mountainous swells were already marching toward California's coast. Wave models predicted twenty- to thirty-foot faces by the time the swells hit Mavericks. Tides and weather were double-checked, and the call was made: the contest's first heat would start at daybreak, January 20.

Big-wave riders had been watching the wave models, too, with gear packed. Within hours of the announcement, the twenty-four invited contestants boarded flights bound for San Francisco from Africa, South America, Europe, Mexico, and Hawaii. They've done this before. In San Francisco, they loaded their long slender boards (*guns*) into cargo vans, along with bags stuffed with fins, wax, leashes, emergency flotation devices, and wetsuits designed for fifty-degree water. Then, like me, they made the forty-five-minute drive to the sleepy town of Half Moon Bay.

Having grown up surfing at Malibu Pier, I've known about Mavericks since it was first publicized in the early 1990s. Before that, big-wave riding was limited to a handful of bold surfers huddled once or twice a year in the lineup at Waimea on Oahu's north shore—native Hawaiians were likely surfing Waimea for centuries. As big-wave surfers grew in number and boldness, they went looking for other breaks. They found Jaws on Maui, Ghost Trees south of Mavericks, Todos Santos in Mexico, Teahupoo in Tahiti, Dungeons in South Africa.

In the mid-1990s a group of surfers left Los Angeles on a boat bound for Cortes Bank, a submerged island a hundred miles offshore, chasing rumors of eighty- to ninety-foot waves. One of the mission's members, *Surfer Magazine's* Bill Sharp, wrote, "It's the only time I filled out a will before a surf trip."

These breaks—and a handful of others that have been discovered since—were either unknown to the surf community or unsurfed thirty years ago. Surfers who seek them out are a different sort, involved in a different sport than those who paddle out at Malibu Pier. As at hundreds of West Coast breaks stretching from Chile to Alaska, the waves at Malibu average three to four feet and peel—"crumble," some would say—benignly across a rocky point. Not so with breaks like Mavericks. These waves don't crumble. They rise suddenly from deep water like a threatened cobra, pitch forward, and break from top to bottom over a shallow reef. These waves are *heavy*. They pack so much force that if a surfer falls, he is often thrown to the bottom and held there for minutes. At best, the experience is likened to being a rag doll tossed in a washing machine's spin cycle. Downed surfers can lose their sense of direction—which way is up and which way is down—and struggle to the surface with only enough time to grab a mouthful of air before being buried under the next whitewater mass.

Most big-wave riders hire a rescue team that stands by on jet skis. A downed surfer is hard to spot amid the turmoil, but his location is sometimes marked by a "tombstoning" board, which stands upright while pulled down by a leash attached to the surfer below. If possible, a rescuer will rush in and pull him out before another wave hits or before the exhausted soul is dashed against the rocks.

Rescue operators are highly trained and put themselves at great risk to save lives, but they're not always successful. In December 1994 surfers watched in horror as Mark Foo, one of the world's most competent big-wave riders, drowned at Mavericks. In 2011 Sion Molosky, a twenty-four-year-old professional surfer, was also lost there.

Greg Long nearly met the same fate last December at Cortes Bank. He and a team chartered a boat to take them out between two large storm fronts. Long was thrown down the face of a sixty-foot wave and slammed against the reef. "I barely made it up for air," he told me. "When I surfaced, another wave was on top of me. I took three set waves on the head and finally blacked out." Later, in an interview at his San Clemente home, he would tell me how the experience changed his life.

Greg Long makes a bottom turn on a Todos Santos monster, December 2005.

I had met Long a couple of years earlier when I learned he was a member of the only surf team granted permission by the Chinese government to ride the Qiantang tidal bore. He and a few other surfers timed their trip to coincide with the fall 2009 spring tides. By the time I read about it in *Surfer's Journal*, I had already been to China to see the bore and had moved on to studying Newton's equilibrium theory.

One of the things I'd learned is that inconsistencies in Newton's theory were surfacing even before it was published. The tide, for example, seemed to behave more like a wave than a bulge. As wave studies progressed through the eighteenth and nineteenth centuries, scientists began to view the tide as a long wave circling the planet—at the speed of a modern jet. After Newton, the next major tide theory, called the *dynamic* or *progressive wave theory*, emerged from this view.

I also learned that a tide wave and a Mavericks wave are more alike than not. It's hard to get our mind around this because we experience them so differently. A Mavericks wave appears on the horizon, rises up, and breaks on shore, all in a matter of seconds. It's a familiar shape—what most of us expect a wave to look like. A tide wave travels ten times the speed of a Mavericks wave, but it's so long—ten thousand miles from crest to crest—that it seems to pass slowly. In fact, it passes so slowly—half a day from crest to crest (high tide to high tide)—that we don't recognize its wave shape.

After reading the story of Greg Long's encounter with the Silver Dragon (where the tide *does* have a familiar wave form), I called him. When he told me he was a competitor, I asked if I could meet him at the next contest. "Sure," he said, "I'll try to get you on a boat." And he did. I would be joining him and a few other competitors, including his brother Rusty, on the press boat, along with photographers and reporters from *Sports Illustrated* and ESPN. We would be positioned alongside the judge's boat, as close to the waves as we could safely get. For me, this was a dream—to see not only the surf contest close up but also what Mavericks could teach me about waves.

~~

On the morning of the contest, I meet Long at the Pillar Point Marina at 5:00 a.m. He's tall and sinewy, with brown hair and eyes. Twenty-nine now, he's been surfing professionally since seventeen, when he won the 2001 National Men's Open Title. Although he seems composed, I know this is his first big wave contest since the Cortes Bank accident less than a month ago. "I paddled out last night," he tells me. "The whole Cortes experience rushed in. I was emotional."

"How are you now?" I ask. "Are you ready?"

"Yeah, I'll be fine. I do this so often that I've learned not to let these feelings carry me away. This sport teaches you to stay in the moment."

It's still dark as we descend into the marina. Thick, tangy air settles among the clutter of masts and fishing gear. A foghorn groans offshore, a constant reminder of a waiting ocean. Most boats rest silently, their mooring lines limp in the predawn calm. The few boats preparing for the day are pooled in light while crews, mostly twentysomethings in jeans, sneakers, and hoodies, busily stow gear and supplies: dry bags in neat rows aft of the pilothouse, surfboards athwartships at the stern, sandwiches and water in the galley. Everything must be lashed down, ready for rough seas.

Our boat's motors are idling as Long and I approach. Rob, the owner, greets us pleasantly but briefly. He slips me two seasickness pills and asks that I sign a waiver releasing him from liability. I do that and down the pills while Long meets with his team to discuss the day's strategy.

Soon we're untying lines and easing out of the harbor, along with four or five other boats. The sky pinkens, revealing details of the marina, Half Moon Bay, and the dry California hills beyond. I once read that when looking at waves from afar, you can tell they are big if they appear to be moving in slow motion. That's exactly the way the waves look at Mavericks as we round the outermost jetty.

〜

Because waves are complicated and elusive, knowledge about them came slowly. In the fourth century BCE, Aristotle noticed that large ocean waves traveled beyond their stormy birthplace, even after the wind had died. Stories of enormous rogue waves arising from nowhere and destroying boats at sea have been part of lore since antiquity. As late as the mid-eighteenth century, some natural philosophers believed that waves were caused by fermentation, which made water swell and explained why waves arrived before a storm.

Leonardo da Vinci, in the fifteenth century, studied waves in their simplest form. Tossing a rock in a still pond, he watched as ripples radiated in all directions. This was simple enough, but eventually the small waves hit the pond's edge and reflected back in the opposite direction. These passed through the initial waves, sometimes combining with them to form larger waves, sometimes canceling them out, and sometimes forming *standing* or *stationary* waves that rose and fell in place. Da Vinci was befuddled—as were scientists for the next five centuries.

Newton hinted that the tide had wavelike qualities. While he was working on the *Principia Mathematica*, his friend Edmund Halley told him about a strange tide at the Red River in the Gulf of Tonkin (present-day Vietnam). In 1678 Francis Davenport, an Englishman who was surveying the Red River for the East India Company, reported that the Tonkin tides rose and fell only once a day, not twice as they did in Britain. "I must needs confess it different," he wrote in a letter to the Royal Society, "from all that ever I observ'd in any other Port." Davenport also noticed that Tonkin's highest tides didn't coincide with syzygy (new or full moon). The news shocked English philosophers, who were convinced that all tides behaved like British tides. Halley called it "wonderful and surprising, in that it seems different in all its circumstances from the *general rule*."

Newton may not have called it wonderful, since it threatened to undermine his new theory. He added a long paragraph in the *Principia* speculating that the anomaly was due to two separate tides coming from different directions. If these tides were *out of phase*, meaning one took twelve hours to reach the Red River and the other took six, they would interfere with each other. The interference might flatten the twice-daily tide and accentuate the once-daily tide. The interference might also stall high tide's arrival to an extent that it no longer coincided with syzygy. The Tonkin phenomenon was much like what Leonardo da Vinci observed on the pond, where waves of different sizes and phases combine to hatch new forms.

Newton didn't call the Tonkin tide *a wave*, nor did he refer to its confusing interaction as *wave interference*, but he implied that this is what was going on. Years later, his speculations were proven essentially true. Meanwhile, other anomalies were showing up. Sailors, in fact, had been discovering idiosyncratic tides ever since they acquired the skill and nerve to set sail for faraway lands.

The Age of Discovery, which began in the fifteenth century and lasted several hundred years, saw an explosion of worldwide sailing expeditions. Motivated by rumors of undiscovered lands, faster and safer trade routes, and the promise of wealth, hundreds of boats took to the oceans. Most maritime nations—Italy, India, the Netherlands, China, South Africa—were engaged, but explorers from Britain, Spain, and Portugal were the most ambitious. Christopher Columbus crossed the Atlantic and bumped into the Bahamas in 1492. Ferdinand Magellan set off from Spain in 1519; one of his five ships limped back three years later, completing the first global circumnavigation. Sir Francis Drake followed, then Loyola, then others.

Because accurate charts and tide tables were nonexistent, these voyagers had to learn the hard way. Boats were lost—most not at sea, but on reefs as they tried to enter unknown ports. Hard-won knowledge of practical navigation—how to take advantage of wind and weather, the depth of port entrances, the behavior of tides and currents—was coveted. Details were meticulously entered in the ship's log and guarded as trade secrets, for commercial and military reasons. Most of these explorers had little interest in tide theory; they just wanted a reliable chart for safe navigation. Yet the farther they ventured, the more baffling the tide picture became.

The tides in Europe occur twice daily, with very little inequality. In other words, if high tide arrives at Liverpool at 1:00 p.m. today, another high tide of equal height will arrive roughly twelve hours and twenty-five minutes later. And because Europe's tides follow the moon, tomorrow's tides will arrive about fifty minutes later than today's. This was all empirically understood during the Age of Discovery, and everyone—both sailors and theorists—assumed this general rule applied everywhere.

Exceptions kept piling up, however. In the Persian Gulf, Philippines, and parts of China and Australia, sailors found once-daily, not twice-daily, tides. In the Gulf of Mexico they found tides that seesawed weekly between once- and twice-daily. In Tahiti the tides seemed to follow the sun, not the moon, arriving like clockwork at noon and midnight instead of slipping fifty minutes later each day. Most of the Pacific had tides with strong and variable inequality, which meant that during a few days each month, there was a large difference in height between the two daily high tides as well as the two daily low tides (see figure on pages 136 and 137).

The general rule was unraveling at home, too. The Solent inside the Isle of Wight on the English Channel had a *double high tide*, and so did Le Havre on the channel's French side. Britain's port of Portland and a few Dutch ports had *double low tides*. And whereas most places had one or two tides a day, Courtown on the Irish Sea had four.

All this pointed to something amiss in Newton's equilibrium theory, which essentially assumed that tides behaved the same everywhere. Newton can be forgiven, since his focus was primarily on what happens in the heavens, not on earth. He wanted to show how the movements of celestial bodies, coupled with that mysterious force called gravity, could stir the oceans. For his purposes, he stripped the earth of its complexities, arrested its orbital motion, jettisoned its continents, and covered the whole thing with deep water. On that blue planet, he was able to demonstrate how the heavenly forces could raise two humps, one pulled by gravity toward the moon and sun and one "fleeing" by centrifugal force on the earth's opposite side. These humps—what we experience as high tide—were free to follow the moon and sun wherever they roamed. It was genius. But it was only half the story.

When the real earth, peppered as it is with landmasses, islands, and continental shelves, is reintroduced to Newton's simplified model, the scene completely changes. Suddenly the watery humps that peaceably followed the moon and sun are blocked. As on da Vinci's pond, they bump into land, reflect back, squeeze through narrow passages, and drag on shallow bottoms. Some of them are *in phase* and some *out of phase*. Some are *standing* and others are *progressive*. To further complicate matters, they do all this on an orb spinning up to a thousand miles per hour.

If the study of tides were divided in two parts—astronomy (what happens *up there*) and fluid dynamics (what happens *down here*)—it could be said that Newton got the first part right. In fact, he got the first part so right that the equilibrium theory continues to be used as an idealized tide model. The second part, what happens *down here*, is unimaginably messy. Scientists are still working it out.

By the early eighteenth century, it was dawning on theorists that there were holes in Newton's theory. With the hope of uncovering a discernible (and predictable) pattern, they called for long-term observations. In 1701 the Académie Royale des Sciences of Paris sponsored observations at Dunkerque, Le Havre, and Brest on the north coast, and within fifteen years they had data (times and heights for high and low tide) for all major French ports. In 1738 the Paris Académie, still unsatisfied with the progress of tide theory, announced a contest for the best essay on "the flood and ebb of the sea." The prize was shared two years later by scientists from four countries. One, a Frenchman, rejected Newton's theory in favor of Descartes's concept of a soupy ether pressing against the oceans. The other three winners were Swiss, Scottish, and Dutch, all building on Newton's equilibrium theory. Even with these developments, tide prediction tables, which were produced privately using secret methods, were still largely inaccurate.

By midcentury, observations were initiated in England at London, Liverpool, Bristol, and Plymouth. In those years, water levels were read on a vertical staff, much like a yardstick, lashed to a dock piling. Astronomers, shipwrights, and dock masters volunteered for the job, which was done more or less hourly, day and night (the first auto-recording gauge wasn't invented until 1830).

(*Overleaf*) A sampling of different tides. Note Galveston's *standing high tide* at quarter moon; Tahiti's consistent noon and midnight high tide; New York's successive highs and lows of near equal height (semidiurnal equality); Los Angeles's *mixed tides*—successive highs and lows, sometimes equal and sometimes not; and Southampton's *double high tide*.

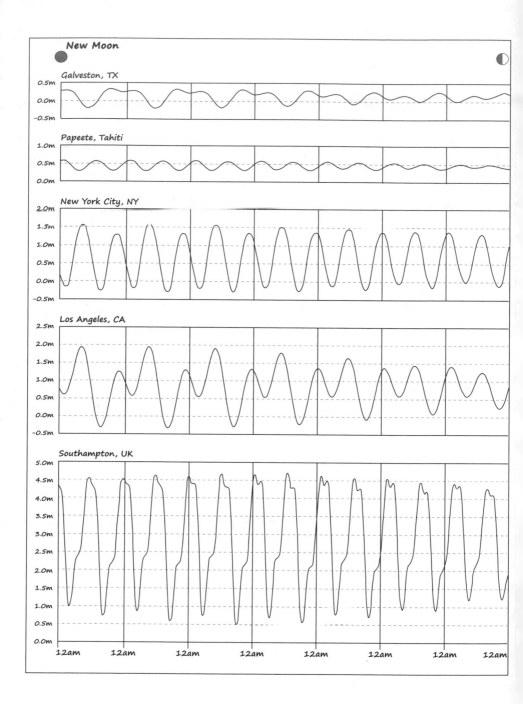

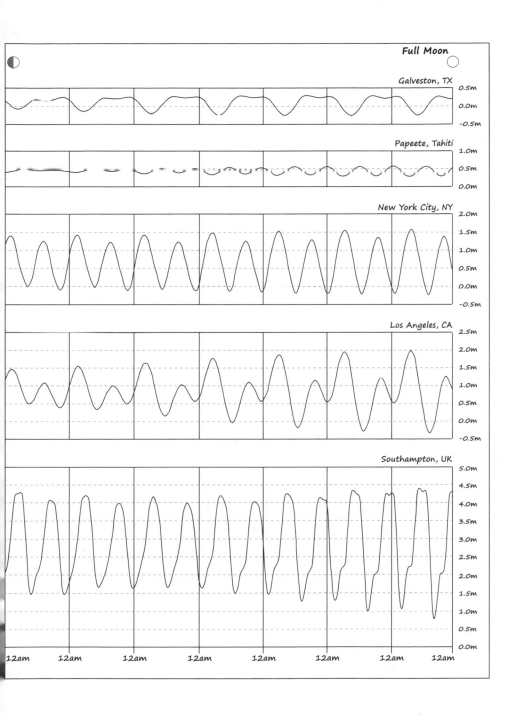

Among other things, these initial observations revealed that tides progressed wavelike, just as the English monk, the Venerable Bede, had suspected a thousand years earlier. "Those who live to the north of me," Bede had observed from his monastery overlooking the Northumbrian coast, "will see every sea tide both begin and end much earlier than I do, while indeed those to the south will see it much later."

The French observations confirmed that if high tide arrived at Brest at 2:00 p.m., it would arrive at Le Havre six hours later and at Dunkerque three hours after that. Low tide would follow the same pattern: arriving first at Le Havre, and finally at Dunkerque.

By the latter half of the eighteenth century, waves had become the central focus of tide studies. Little was known about them, but enough to glimpse how their behavior might unlock the tide's mysterious workings in the real ocean. This new direction, called the *dynamic* or *progressive wave* theory, was a major stepping-stone between Newton's theory and the current *harmonic* theory. In it, Newton's watery bulges were visualized as large, long waves racing at 450 miles per hour around the Southern Ocean, where bothersome continents or shallow water wouldn't interrupt them. From there, they spread northward into the Atlantic, Indian, and Pacific Oceans.

≈

Our boat pitches and rolls in the swell as we approach Mavericks. I'm drowsy from the Dramamine, but it beats being sick. The sun is up, and the first six surfers paddle out. Rescue teams zip here and there on jet skis. The San Mateo County sheriff's boat stands by, as does a medical emergency boat. A couple of kayakers and a few local surfers have paddled out from the marina. As the day wears on, thirty or forty boats bob in a cluster, well away from the impact zone but as close to the waves as the officials will allow. Occasionally a gray whale surfaces offshore, its misty exhale hanging in the light breeze.

The waves come in sets of three or four, warping from the horizon like blown glass. Even at this close range, they seem to build in slow motion. For most of their journey, they've been able to speed along in deep water without feeling the bottom. Now, encountering the reef, they're transformed. They drag in the shallows and steepen to a point of instability. As the bottom of the wave slows, the upper lip rushes forward, silent and trembling, until the whole thick slab collapses. The wave's leading edge hits the water below with a deafening crack, as if the ocean were concrete. Masses of water bounce into the air. Seconds later, a whitewater avalanche emerges from the chaos and rumbles toward shore. During sets, the impact zone—where the waves break— is draped in a haunting mist.

I watch several competitors paddle for these watery mountains. Their performance is judged by the biggest wave, the most critical drop, making it (not falling), and sportsmanship-like conduct (not putting themselves or others in undue danger). Each heat is forty-five minutes, with the top six competing in the final heat. These six often agree to split the prize, in this case $50,000, a tradition inspired by Long when he won the contest in 2008.

During a later heat, I watch through binoculars as Long goes for a large set wave. He puts his head down and paddles furiously. The wave comes up from underneath, and only at the very last moment, high on the lip, does he catch it. He stands up and hangs for a second or two. Then, as if the wave has decided to allow him in, he races down the face and makes a long, arcing bottom turn.

"It's all in the drop," he tells me later. "These waves are moving about thirty-five miles per hour. It's hard to paddle fast enough to catch them. When you get in, you're usually high on the lip. It's a critical moment. You hang there weightlessly, looking down a cliff and trying to set your position for the drop. The wave is made or lost in those first moments, which feel like eternity. If you make the drop and the bottom turn, the rest is icing. If you don't make the drop—if you get hung up on the lip or don't get positioned on the board right—that's the end. You're thrown down the face, and a wall of water three stories high unloads on you."

At twenty-plus feet, the waves today are plenty big, but Mavericks has seen larger. In November 2001 they topped sixty-five to seventy feet. Brazilian Carlos Burle surfed a record-breaking sixty-eight-foot wave that day. His record has been broken several times since—at Cortes Bank, Jaws, and most recently Nazare, Portugal, where in 2013 and 2014 waves up to a hundred feet may have been ridden. It's difficult to confirm the height of these waves, since there is no consistent and verifiable method of measurement.

When big wave surfing was discovered in the 1990s, it was thought that these monsters were moving too fast to paddle into. Surfers were towed by jet skis and whipped into the wave well before it began to break. By doing this, they avoided the treachery of hanging up on the lip as well as making the drop.

In 2008 Long and a small group of friends wanted to find out if they could paddle into these giants. They studied wave dynamics and tested new equipment: longer, narrower, heavier boards, designed for paddling faster. "Eventually, we found out it can be done," he says, "but it requires a different set of skills than tow-in. You have to know more about the wave, how it's moving, where the peak is forming, how much current is coming up the face, where it's going to break. The takeoff zone is narrow—if you're too far one way or the other, it could mean missing the wave or getting hurled. And your only source of power is your hands."

Long doesn't know the size limit for waves that surfers can paddle into, but he wants to find out. Records are broken frequently. Shawn Dollar paddled into a fifty-five-footer at Mavericks in 2010, and Shane Dorian caught a fifty-seven-footer at Jaws a year later. Shawn Dollar holds the current record with a sixty-one-footer at Cortes Bank in 2012. "There's always the hope that there's a bigger wave out there," says Long. "You never know when it's going to happen, or where. The biggest waves we've ever seen could show up ten days from now—or ten years from now. That day *will* happen, and you can bet there'll be surfers ready for it."

～

These big waves are almost always born in offshore storms, traveling in organized sets across vast distances. As wind-generated waves, they're common and conspicuous. Benjamin Franklin, who was fascinated by the ocean, wrote in 1774: "Air in motion, which is wind, in passing over the smooth surface of water, may rub, as it were, upon that surface, and raise it into wrinkles, which if the wind continues, are the elements of future waves."

These first wrinkles are what we now call *cat's paws* or *capillary waves*. They have a *period* (the time between the passage of two crests) of fractions of a second. If the wind blows hard enough, long enough, and over a large enough distance (fetch), these wrinkles build into waves, at first a foot or two high and eventually into a fully developed sea of twenty-foot-plus waves.

These waves leave their stormy nursery and traverse the oceans in organized sets, or swell trains, with periods of fifteen to twenty seconds. Surfers and sailors much prefer these over local wind waves, which tend to be short, steep, and "confused."

All wind waves—even a hundred-foot rogue—begin as a wrinkle. With life spans up to several days, they usually expel their last breath when they break on a reef or beach. Once they break, their life is over.

Tsunamis, storm surges, and tides are waves, too, but with different characteristics. Tsunamis (Japanese for "harbor waves") are spawned by seismic events, volcanic eruptions, landslides, or earthquakes; a meteorite, if it hit the ocean, could also cause a tsunami. They're not as common as wind waves, but they're more common than we realize. Catastrophic tsunamis occur about twice a decade—mostly in the Pacific Ocean—but smaller ones happen several times a year. In the open ocean, a tsunami, like the tide, is a long, low wave that travels about 450 miles per hour. With a period of thirty to ninety minutes, these waves are so long and low—just a foot or two high at sea—that they're virtually undetectable by boats offshore. Yet when tsunamis approach a coast and begin to feel the bottom, they slow down and steepen just like a Mavericks wave, sometimes towering to a hundred feet and causing horrible damage and loss of life.

A storm surge is created by a combination of low barometric pressure and high winds associated with a storm front. Essentially it's a single hump of water—as high as forty feet—traveling directly below a storm. When the storm makes landfall, so does the surge. Like tsunamis, storm surges occur far less frequently than wind waves, but they're often memorable. Combined with a large tide, as in the case of Hurricanes Katrina and Sandy, they can be catastrophic. Katrina's storm surge, which hit the Gulf Coast in 2005, was twenty-two feet, the largest in U.S. history (Katrina's total tide-plus-storm surge was a record thirty-four feet). Sandy's, which made landfall in New York and New Jersey in 2012, was nine feet.

Some waves feel the bottom and some don't. If they do, they're *shallow-water* waves; if they don't, they're *deep-water* waves. Whether they're one or the other is determined by the wave's *length* (distance between crests) relative to water depth. If a wave is traveling in water with a depth of less than half its wavelength, it's a shallow-water wave. In other words, a wave with a length of ten feet is a shallow-water wave if it's traveling in a depth of five feet or less.

With a wavelength of eighty to a hundred miles, a tsunami is a shallow-water wave, even as it crosses the deepest ocean basins. The oceans would have to be forty to fifty miles deep to allow it to move without feeling the bottom.

Tides are shallow-water waves, too. With a period of just over twelve hours and a wavelength of ten thousand miles, they would need an ocean five thousand miles deep to race at full speed around the earth. The ocean's average depth, however, is only three miles. So, like a tsunami, the tide constantly drags its feet on the sea's bottom, adding layers of complexity to its shape and timing.

There's another important difference between these waves. Wind waves, tsunamis, and storm surges have a beginning and an end. They're born of a single event—a puff of wind, a storm, or an earthquake. As *free waves*, they travel away from the event without receiving further energy from it. Although they can traverse enormous distances, they eventually perish for loss of energy. A free wave behaves like a child on a swing whose parent gives him just one push and walks away. With no further pushes, the child sweeps back and forth in smaller and smaller arcs until the initial energy dies out and the swing stops.

Although the tide can act like a free wave at times, it's primarily *forced*, not free. That means its "parents"—the moon and sun—push continuously. They never walk away—not today, not yesterday, not tomorrow. Consequently, a tide wave, unlike a wind wave or storm surge, has no beginning or end. It never was a wrinkle. It doesn't break and die.

〜

From my perch on the bow of Rob's boat, I watch another set roll in. The outermost wave rears up a third again as high as the others, perhaps the largest and steepest we've seen today. Peter Mel catches it and screams down the face, his arms stretched winglike for balance. The wave's lip feathers; a white plume blows off the back side. Through binoculars, I can see the wave sucking water off the reef and up its steepening face. The lip reaches skyward and arcs forward, creating a classic almond-shaped hollow. It's a gorgeous wave, monstrous and silky, breaking cleanly across the reef. Mel makes the drop and bottom turn, then stalls on the shoulder, positioning himself for the barrel.

In a moment he's gone, folded into the wave. Has he fallen? There's no way of knowing. The wave's open end is choked with spray. If he's in there, he's crouched low, head ducked. His right hand is trailing, touching the wave, feeling its pulse. Being in the barrel is the most intimate experience of a wave. Every surfer dreams of it. Here, mind, body, and ocean are caught in the most exquisite and ferocious balance. Time stops. The kinetic energy is wound so tightly that one wrong move, as slight as catching a finger on the water, snuffs out the experience.

Our eyes are riveted on the wave's leading edge, which huffs and snorts like a heaving animal. Just when it seems impossible that Mel could still be standing, he's spit from the barrel.

We're stunned, relieved. Everyone, including the other competitors, lets out a cathartic cheer. I overhear a judge say, "That might be the wave of the day." No one would disagree. It's a wave few of us will forget, especially Mel.

Some waves are like that: unforgettable. Of the thousands—millions—that come and go, a few stand out. Surfers like Mel and Long remember waves that gave them the best—and the worst—rides. Long, I'm sure, won't soon forget the wave that almost killed him at Cortes Bank.

Sailors, too, remember certain waves. I won't soon forget the tide wave that almost sank *Crusader* at Kalinin Bay in southeast Alaska many years ago. Nor will I forget a wave I encountered while making a 1,500-mile passage from Florida to the Virgin Islands in the early 1980s. A friend and I had been sailing in a storm for several days. We had taken all the sails down, but my twenty-six-foot sloop was still "sailing" at hull-speed under bare poles, blown by forty-knot winds. The waves were more than twenty-five feet, each one coming from behind, lifting my small sloop's stern and passing under. But one wave reared up perhaps one and a half times as high as the rest. I could see its blackened face looming above the others, smothering the horizon.

I had only one thought: if this monster breaks, it will be the end of us.

I yelled to my friend in the cabin to brace himself as our little boat began to climb—up and up and up. Then, suddenly, we rushed forward and surfed down the face. At the bottom, we broached to starboard and were overwhelmed by whitewater that snapped lashings and carried away whisker poles, extra gas cans, life jackets, and the self-steering vane. I would have been carried away too if I hadn't been attached with a life harness. After the wave passed, my friend and I bailed several hundred gallons from the cabin. We were shaken and bruised—but grateful to be alive.

Waves don't have to be big or menacing to catch our attention. Most anyone who has spent time on or near the ocean remembers a wave or two, perhaps one we admired for its shape or grace as we strolled the beach or one that caught us by surprise and tumbled us amid the shore break as a child.

Traditional Polynesians made an art out of remembering waves. While most of us see the ocean's surface as a blur of motion, they saw it as a pattern to navigate by. When miles at sea, they could tell by the crisscrossing swell behavior exactly where they were and the course they needed to steer—and for how long—to get home. This knowledge didn't come quickly, but was acquired over thousands of years of practice.

They even named their waves: *rilib* was the dominant northeast swell, *bungdockerick* was a southwest swell, and *buoj* was the intersection of east and west swells. How these swells bend and twist around islands was mapped on stick charts and memorized. At sea, a good navigator could lie in the boat's hull and know his whereabouts just by sensing the rolling motion of the swell. Navigators in these cultures, identified with a "gift" at childhood, served as apprentices for years before taking charge of a seagoing boat.

⁓

Greg Long studies waves as diligently as a gifted Polynesian navigator. "Each break," he tells me, "has a personality that's determined by the bathymetry—water depth, shape of bottom, size of reef, and so on. Bathymetry never changes. It can focus waves or diffuse them. Yet any given wave on any given day, even hour to hour, is unique depending on the swell size and direction, wind, and tide. It's always shifting. Onshore winds flatten waves, and offshore winds stand them up. A difference of a few degrees in swell direction can influence how a wave will break too. At Mavericks, the wave is predictable and pretty clean. It usually breaks in the same spot every time. At Jaws and Dungeons and Cortes, the wave can stand up and break at different places on the reef, so you're always on edge, especially if you're trying to paddle in."

Tides also play a role in shaping surf breaks. The flood can push a swell in; ebb can flatten it. Some breaks only "show" for an hour or two during the right tide or when just enough water is over the reef. Some breaks are too dangerous to surf at low tide. "I had one of my biggest accidents at a low tide at Mavericks," says Long, "and learned not to do that again. If there's too little water over the reef, the wave stands straight up and breaks like one large slab. I fell on one of those and was swept across the reef into deep water. I probably went down fifty feet. It was like going over an underwater waterfall. I went down so fast that I broke an eardrum. That turned into a two-wave hold-down. I've experienced about six two-wave hold-downs in my life, and four were at Mavericks."

When not surfing, Long works out five hours a day. His routine includes yoga, swimming, mountain biking, and breath-holding exercises. "My father was a lifeguard and taught my brother and me respect for the ocean," Long says. "I take that very seriously. People think I'm crazy, but I'm always rehearsing worst-case scenarios in my mind, trying to figure out how to prepare for them." Long eats mostly raw food, designs surfboards, and tests safety equipment. "I have a team I like to surf with—guys I trust. We rehearse different rescue scenarios. I won't go out at a break like Cortes without a six-man safety team."

Long also studies wave models. On his computer, he keeps seven years' worth of wave, storm, and tide statistics. "I watch almost all the big oceans—the Atlantic, Indian, North and South Pacific—daily," he says. "When I see a storm crop up, I can superimpose my statistics and predict where and when the waves will hit and how large they'll be."

Among elite surfers, Long stands out for his interest in and knowledge of waves. Surf writer Brad Melekian remembers an afternoon phone call from Long a few years ago when the two were supposed to meet at a yet-to-be-determined break the next day. "First thing I hear," Melekian says, "Greg's on Highway 1, driving to Mavericks. He pulls over, takes out his laptop, checks the buoys. Calls and tells me he's just booked a boat for Cortes. Then calls back fifteen minutes later:

'We're going to Shark Park' [near Santa Barbara]. Fifteen minutes later: 'Mavs.' An hour later: 'Okay, Todos.' And then finally it's back to Mavs. That's what he does. He's processing and analyzing nonstop."

～

A handful of eighteenth-century natural philosophers were processing and analyzing waves too, most notably William Whewell, George Airy, and Pierre-Simon Laplace. In the context of the progressive wave theory, understanding wave behavior was the next best thing to understanding the tides.

Whewell (1794–1866), an Anglican priest and master of Trinity College, Cambridge, pushed for more field observations. When his attention turned to the sea, he was thirty-six, already an accomplished scholar of mineralogy, architecture, history, astronomy, and poetry. He's perhaps best known for his writings on the philosophy of science, but his interests were eclectic. He wrote sermons, lectured on Gothic architecture, translated Johann Goethe, tutored Alfred, Lord Tennyson, and coined the terms *ion, anode, cathode,* and *physicist.* In 1833, after the poet Samuel Taylor Coleridge complained that "natural philosopher" was the wrong term for a person digging in fossil pits or performing technical experiments, Whewell offered the word *scientist.* If "philosopher" was "too wide and lofty a term," Whewell said, "by analogy with *artist,* we may form *scientist.*" And so it was.

Prior to Whewell's involvement with tide studies, John Lubbock, a fellow Trinity mathematician and astronomer, dug into the Liverpool and London tide records, the oldest sustained observations in Britain, in search of a long-term pattern that could be used in developing a reliable tide prediction method. Prediction at that time was a matter of practical knowledge and relied on the fact that every port had a unique and unchanging relationship between the moon's passage overhead and the arrival of high tide. The method was called *Establishment of Port.* In practice, a sailor journeying down Britain's coast knew that the Liverpool harbor entrance, for example, was shallow and only passable at high tide. He also knew that the establishment at Liverpool for that night was plus one hour and twenty minutes. With that knowledge, he would stand off the harbor entrance until the moon reached its zenith, wait one hour and twenty minutes, and sail safely in.

The term *Establishment of Port* was coined by Whewell, but the principle was surely recognized and used by early navigators. In the eighth century, Bede wrote: "In a given region the moon always maintains whatever bond of union with the sea it once formed."

Lubbock was less interested in Establishment of Port and more keen on the secretive calculations behind the tide tables produced by private parties in Liverpool and London. The Liverpool tables had been continuously produced by one family for fifty years (since 1770). Lubbock eventually won their trust and was able to review their methods, which were loosely based on the work of Daniel Bernoulli, a prizewinner of the 1740 Paris Académie contest.

Lubbock presented his findings at a Royal Society meeting in 1830. Whewell was there and, inspired, soon took over where Lubbock left off. True to his wordsmithing instinct, he called the pursuit *tidology* (the name never stuck), and in the following twenty years he published fifteen tide papers in the *Philosophical Transactions*.

Whewell was more ambitious than Lubbock. He recognized that while his colleague's effort was admirable, it would ultimately reveal only the tide pattern for a single port, such as London or Liverpool. Because the tide varied considerably from port to port, Lubbock's method wouldn't reveal a global pattern. If there were a larger pattern, figured Whewell, it would be revealed only through *simultaneous* observations *at many ports*. By doing this, a line could be drawn connecting the places of high tide at any given hour. Whewell called such lines *co-tidal*.

Whewell couldn't organize a large set of observations on his own, so he appealed to Sir Francis Beaufort, head of the Admiralty's Hydrographic Office. Beaufort, who would develop the Beaufort Scale of wind speed still in use today, responded enthusiastically. While

Whewell wrote instructions for observers, Beaufort ensured the cooperation of more than four hundred Coast Guard stations around the British Isles. For two weeks in July 1834, tide observations were made every fifteen minutes. When the data came back, they revealed an odd picture in the North Sea (then called the German Sea). Whewell proposed a second experiment to confirm the results, but Beaufort was already thinking bigger. He wanted to expand the experiment internationally.

In June 1835 the two men coordinated history's first large-scale international scientific experiment. More than six hundred stations participated in a three-week survey: five hundred in the British Isles, twenty-eight in the United States, twenty-four each in Denmark and Norway, eighteen in the Netherlands, sixteen in France, and a dozen each in Belgium, Spain, and Portugal.

The experiment produced sixteen thousand sets of data. The daunting analysis was undertaken by three clerks from the Admiralty (*computers*, as they were called). It took months. Anxious for the results, Whewell wrote to a friend, "I shall have such a register of the vagaries of the tide-wave for a fortnight as has never before been collected, and, I have no doubt, I shall get some curious results out of it."

He had no idea how curious the results would be. The experiment, he expected, would confirm the tide's northward progression from the southern reaches of the Atlantic, with high water arriving at roughly the same time on opposite coasts. When the tide got to the British Isles, it would pass the English Channel (due to its narrowness) and speed northward around the northern end of Scotland before flooding south into the North Sea.

The picture that emerged from the mass of data, however, showed that the tide's progression in the North Sea was circular, not linear. It looked and functioned like a pinwheel, with a hub at its center and spokes radiating outward. These spokes circled counterclockwise (see figure on next page). The center hub implied a "point of no tide." The farther out on the spoke, the larger the tide. Additionally, it appeared

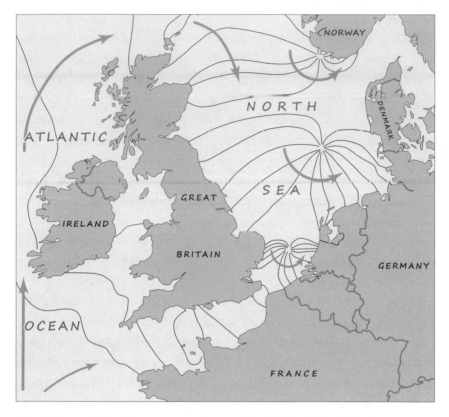

North Sea amphidromic systems, as discovered by William Whewell in 1835.

that the North Sea, as small as it is, had more than one *rotary system*, as Whewell called it, and perhaps as many as three.

The Astronomer Royal, George Airy (1801–1892), ridiculed the notion of rotary systems. He was keen on wave study, too, and had learned by testing in a laboratory channel how waves bend, reflect, stand, interfere, and refract. He figured the North Sea anomaly was explained by *wave interference*, just as Newton had explained the Tonkin tides earlier.

Whewell's discovery was anachronistic, as if he had accidentally opened a box that was not due to be disturbed for another fifty years. What was he to make of these spidery things? If they were in the North Sea, were they in other oceans too? At Whewell's urging, the Admiralty sent out a ship in 1840 to confirm that there was, in fact, little or no change of water level at the "point of no tide."

Whewell wanted more information. Twice he petitioned the Royal Navy to launch a worldwide two-year ocean expedition to "hunt" the tides. When the navy showed no interest, Whewell shifted his focus back to the data he had at hand but met a dead end there too. In 1866, when the results of the International Tide Experiment were being debated and analyzed in Europe, reports were coming from the South Atlantic that the tide wasn't behaving as it should down there, either. Instead of progressing systematically northward, high tide seemed to seesaw in an east-west fashion, from one coast to the other.

Captain Fitzroy, master of the HMS *Beagle* that took Charles Darwin around the world from 1831 to 1836, delivered the most damning news for those still clinging to the progressive wave theory. From years of firsthand experience, he reported that there wasn't a noticeable tide progression along 1,800 miles of Africa's coast, from the Cape of Good Hope to the Congo River, and that the South American coast around the Plata River had almost no tide. Fitzroy proposed that each ocean might have its own tide that sloshed back and forth, just as Galileo had observed in the jugs of water while on a boat in Venice.

George Airy agreed but called the Atlantic tide a *standing* or *stationary* wave. From his experiments, he knew that a standing wave in a channel rose and fell alternately at each end, with a *node* of no motion in the center (like a bathtub full of water when we rock back and forth). If this were the case in the Atlantic, then the tide would rise and fall on the east and west coasts but not in the midocean, which would explain why offshore islands generally have small tides.

Whewell's speculation, not surprisingly, was that the Atlantic tides behaved like a North Sea pinwheel, but he was unable to prove it. In an 1847 lecture to the Royal Society, he admitted, "I do not think it likely that the course of the tide can be rightly represented as a wave traveling S-N. . . . We may much better represent it as a stationary undulation, of which the middle space is between Brazil and Guinea [where] the tides are very small."

⁓

French mathematician Pierre-Simon Laplace (1749–1827) died before seeing the results of the International Tide Experiment, but he had already seen beyond it. Often called the Newton of France, Laplace was distinguished for his genius in mathematical formulations, many of which are still in use today. Like Newton, his attention to tides was fleeting compared to the main body of his work, but it had far-reaching implications. He called the tides "the thorniest problem in astronomy."

In Laplace's five-volume masterpiece, *Mécanique Céleste*, he introduced equations to address the complicated interactions of tide waves on the real earth. He recognized that there was more to the ocean tides than a simple wave progressing around the planet. Instead, he described how each ocean might have its own response to the tide-generating forces and that that response might be defined by many factors, including the size and shape of the basin, the depth of the water, the ruggedness of the bottom, temperature, and so forth. Using calculus and trigonometry, he developed several highly sophisticated equations to account for this, equations that turned out to be nearly impossible to solve without modern-day computers, which wouldn't be in use for another 150 years. He never fully solved them himself. They looked like this:

$$y = -\frac{l}{\sin\theta}\frac{\partial . u\gamma\sin\theta}{\partial\theta} - l\gamma\frac{\partial v}{\partial\varpi},$$

$$\frac{d^2 u}{dt^2} - 2n\frac{dv}{dt}\sin\theta\cos\theta = -g\frac{\partial y}{\partial\theta} + B\Delta + \frac{\partial R}{\partial\theta},$$

$$\frac{d^2 v}{dt^2}\sin^2\theta + 2n\frac{du}{dt}\sin\theta\cos\theta = -g\frac{\partial y}{\partial\varpi} + C\Delta\sin\theta + \frac{\partial R}{\partial\varpi},$$

In my quest to understand the tides, I've sat with many oceanographers who were eager to share their knowledge. Even when I begged them not to, they often turned to equations, insisting on their clarity and elegance. I agree, and admire equations like those above as an expression of the tide, just as a sail's shape is an expression of the wind or a surfboard's shape is an expression of a wave, but I don't have a clue what these equations mean.

Hidden within the unsolved Laplacian equations were the seeds of modern *harmonic tide theory*, which views the oceans as vibrating basins that respond to the influences of the moon and sun. Also hidden in the equations was the solution to global tide prediction, which was not to be deciphered until the late nineteenth century.

Like Laplace's equations, Whewell's concept of rotary tides laid asleep for many years, not to be awakened until the early twentieth century. By then, the Coriolis effect was understood: the earth's rapid rotation makes wind, waves, storms, and even baseballs veer to the right in the northern hemisphere and to the left in the southern hemisphere. Rollin Harris (1863–1918) of the U.S. Coast and Geodetic Survey saw in 1904 that the tides veered, too. In fact, the tides veered so dramatically that they circled around a center point in exactly the same way Whewell had seen in the North Sea. And they were everywhere (see figure on next page).

These circular systems reminded Harris of an ancient Greek naming ceremony called *amphidromia* ("run around"). About seven days after a child was born, family and friends gathered to drink wine, eat octopus and cheese, hold hands, and dance in a circle around the newborn. In some cases, the child was ritualistically carried around the household hearth. Harris, evidently more of a romantic than Whewell, called these pinwheel-like tides *amphidromic systems* or *amphidromes*. Like the Greek ritual, an amphidrome has a center hub where, as Whewell proved, there is little or no tide. The arms or spokes rotate, with the highest tides in each amphidrome occurring farthest from the center hub.

Some—not all—of the world's amphidromic systems.

The Pacific has four large amphidromes. One sits west of Baja, with its long arms sweeping up and down the North and South American coasts. The progressive wave theory, like Newton's theory, was not completely wrong. What wasn't thrown out—and is still part of the tide story today—is that these circling arms are indeed waves traveling at 450 miles per hour.

⁓

The last heat finishes at Mavericks as the late-afternoon sun turns the California hills orange. We sail back to Half Moon Bay for the award ceremony. No one is surprised when Peter Mel wins. Long comes in third, but all six in the final heat split the $50,000 purse. "We agreed to do that at the start of the heat," says Mel, a native of Santa Cruz, "because it takes the pressure off. It usually brings waves, too, and that's what happened today."

In August 2013 I meet up with Long again, at his San Clemente home. The Cortes Bank accident of nine months ago is less raw, but still on his mind. "I knew the waves were going to be big that day," he says, "the offshore buoys were reporting a fourteen- to fifteen-foot swell at nineteen seconds. That would mean up to eighty-foot faces at Cortes." Long assembled a team of trusted surfers and rescue operators. On the evening of December 22 they met at Newport Harbor and loaded their gear and six jet skis aboard a 120-foot charter boat. They left the harbor at 10:00 p.m. and arrived at Cortes early the next morning.

All day, sets rolled in with two waves each. As a rule, Long never paddles for the first or second wave of a multiwave set, for two reasons. First, if he falls, he risks being trapped in the impact zone where the rest of the waves break. Second, Long (like most other big-wave riders) is always hunting for the biggest wave, which is almost never the first wave in the set.

On that day at Cortes, Long took off on an early wave of what turned out to be the day's only five-wave set. He made the drop, but the wave closed out as he finished his bottom turn. Then he fell.

He was thrown onto the reef and buried in thirty feet of whitewater. "When the wave let me go, I swam for the surface," he says, "but just as I was about to get a breath, the next wave broke on top of me and knocked the wind out of me entirely. I was pushed right back to where I had been fifteen seconds ago, with no air in my lungs." Being held down for twenty to thirty seconds, Long explains, is something all big-wave surfers are accustomed to. They train for it. To be buried for twenty-five to forty-five seconds in a two-wave hold-down, however, is highly unusual and, in his words, "terrifying."

But Cortes Bank was different. It was a "perfect storm," says Long. The second wave pushed him down to about fifty feet. "I was completely out of air," he says. "My muscles were spasming. I've never felt that much pain. I also knew my bloodstream was still saturated with oxygen and that the body can usually survive longer than we realize.

"At that point," Long continues, "it became a mental battle. To survive, I knew I had to separate from it mentally. My yoga practice has helped me with that. From a physical side, the exercises have expanded my lung capacity, and on a spiritual side, it's taught me the value of meditation."

Long was still deep underwater when he heard and felt the third wave pass. He'd been down for about a minute and knew he was going to black out. He had a choice: he could black out down there, or he could fight for the surface, where the rescue team would have a better chance of finding him. He grabbed his leash, which luckily was still attached to the board, and climbed. At the same time, he felt himself separating from his body, letting it go.

Just as Long reached the tail of his tombstoning board, he blacked out. A fourth wave washed over before he surfaced, facedown. The rescue team, including two emergency medical technicians, pulled him out and raced him back to the support boat. The Coast Guard was called. For about a minute and forty-five seconds, Long had no pulse. "That part of the experience is less clear," says Long, "but I know I left my body at least a few seconds before I blacked out." Clinically, he had drowned.

Long was coughing blood when he regained consciousness. In such cases—nonfatal drownings—the victim often suffers brain damage from oxygen deprivation. As night fell, a Coast Guard helicopter arrived and flew him to the University of California–San Diego Medical Center. Long was lucky: he was released from the hospital the next day, Christmas Eve, with no further complications—at least physically.

"When I met you at the Mavericks contest in January," Long says, "I was still shaken up. The accident had happened a few weeks earlier. I didn't know what I was doing or even why I was surfing anymore. I was just going through the motions, trying to power through it, by sheer ego, pretending I was okay. But I wasn't. I was terrified. Nothing felt the same."

It seems that most people would want to put a harrowing experience like this behind them—to erase it from memory. Long does the opposite. He keeps circling back to the Cortes incident, reliving every detail through dreams and meditation, in search of more knowledge, more insight, more lessons—a search that is like his thirst to understand waves and the ocean.

"Since then," Long reflects, "I haven't stopped surfing. I've been following nearly every big swell as I have my entire career. Each session I go out, I regain some confidence and comfort. Even with the fear, I have no regrets about the accident. Everything's a lesson—and Cortes was a huge one for me. Now I'm getting through it and living life with a different perspective on big-wave riding—a different set of challenges or rules than the next guy, who's maybe never had a bad wipeout. I don't care anymore who surfs the biggest wave or wins the contests. I only care about passion—about loving what you do. And what I love to do is find and catch big waves."

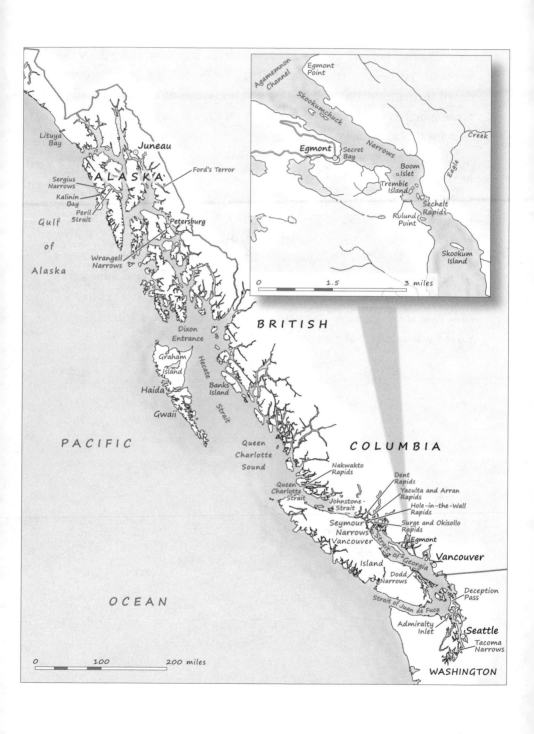

Lituya
Bay

Juneau

ALASKA

Ford's Terror

Sergius
Narrows
Kalinin
Bay
Peril
Strait

Petersburg

Gulf

of

Alaska

Wrangell
Narrows

Dixon
Entrance

Graham
Island

Haida

Banks
Island

Hecate

Gwaii

Strait

PACIFIC

Queen
Charlotte
Sound

BRITISH

COLUMBIA

Nakwakto
Rapids

Dent
Rapids

Queen
Charlotte
Strait

Johnstone
Strait

Yaculta and Arran
Rapids

Hole-in-the-Wall
Rapids

Surge and Okisollo
Rapids

Seymour
Narrows

Egmont

Vancouver

Vancouver

Strait of Georgia

Island

Dodd
Narrows

Deception
Pass

OCEAN

Strait of Juan de Fuca

Admiralty
Inlet

Seattle

Tacoma
Narrows

WASHINGTON

0   100   200 miles

Agamemnon
Channel

Egmont
Point

Skookumchuck

Creek

Egmont

Secret
Bay

Narrows

Eagle

Boom
Islet

Tremble
Island

Sechelt
Rapids

Roland
Point

Skookum
Island

0   1.5   3 miles

6 Fast Water

How Tidal Currents Slow the Earth and Bend Time

Time and Tide stayeth for no man.
—*Richard Brathwaite, 1630*

~ At one in the morning the tide is high and a full moon is loose in the meadow. Awakened, I wrap a blanket over my shoulders and follow the flood of light and shadow into my cabin's living room. Bright as a car's headlamp, the moon's milky rays spill over a nautical chart I laid out last night. Beside it is a stack of reference books: a *Nautical Almanac*, *Sailing Directions*, a *Coast Pilot*. Bags of rice, flour, oatmeal, and coffee clutter the kitchen countertops. Tomorrow I'll stow all this on my schooner, *Crusader*, and set sail on a four-day trip north into British Columbia.

I'm not sleepy, so I gaze over the chart, noting the sailing routes and myriad numbers indicating water depths, navigation aids, and compass directions. Channels and reefs are marked with buoys. Some of these, the chart indicates, are lit at night with red, white, or green lights. Some are "quick flashing," others flash every two, three, or four seconds. Sailors count between flashes—one one-thousand, two one-thousand—to confirm which light is which. The difference could mean making it safely into port or going aground. The chart also notes the location of shoals and rocks that bare at low tide, prominent headlands and lighthouses, and, in large purple swaths, shipping lanes. The fine print reminds sailors that depths are in fathoms, equal to six feet, and distances are in nautical miles, equal to 1.15 statute miles.

As I unfold the chart, I see hundreds of skinny passages between Vancouver Island and mainland British Columbia, as if the two were once a single loaf of bread that has been pulled mostly apart. The main waterway—the Inside Passage—hugs Vancouver Island's east coast, snaking six hundred miles through rain-soaked, evergreen islands to Alaska. Countless veins wander off through steep-sided valleys far into the mainland's interior.

All this was under mile-thick ice ten thousand years ago. As the planet warmed, the heavy ice retreated, scraping and clawing across the landscape, leaving the craggy coast we know today. The sea followed, filling the deepest-cut valleys. Where the valleys narrowed, the retreating ice left large piles of gravel and rock, called sills. Over these, present-day water depths jump from hundreds of feet to just twenty or thirty, then plummet again. The result is a channel that's pinched from all sides—both shores and the bottom—like an hourglass.

These narrow channels create havoc for the tide, which floods every small vein of the Inland Passage in just six hours, then rushes out in the same amount of time, clearing the way for the next cycle. In a deep, island-free ocean, this kind of speed might be possible, but never in a shallow, island-strewn, pinched-up coast. Here, the tide behaves like rush hour traffic when five freeway lanes are suddenly reduced to one. An important difference, however, is that automobile traffic can stop and wait. The tide can't. Won't. In its inexorable momentum, it stumbles over the shallowing bottom, plows into islands, piles up at the head of snaky corridors, and finally, at hourglass narrows, bursts into a torrent of whitewater and sound. The equivalent on a freeway would be if the cars piled on top of each other and speeded up to all get through a single freeway lane at the same time.

I put my finger on the chart and trace northward, counting the narrows. In the south, I slip over Tacoma Narrows and Admiralty Inlet, stopping just long enough at Deception Pass to read the warnings: "rips and boils . . . should not be attempted by a small boat during maximum tidal currents." In British Columbia, my finger skips across dozens more rapids, slowing only at the largest ones: Seymour, Yaculta, Surge, Sechelt, Hole-in-the-Wall, Arron, Okisollo, Nakwakto. In Alaska, I find the big rapids of Peril Straits, Sergius Narrows, and Ford's Terror.

Some of these narrows have currents of two or three knots, so they are navigable no matter which way the tide is going—although for fuel economy most sailors choose to travel with, not against, the tide. A few, like Sechelt, Sergius, and Nakwakto, have currents that reach world-record speeds of more than eighteen knots. A prudent sailor wouldn't think of entering one of these places during maximum current. Nor would the captain of a large ship. In Alaska, the 400-foot-long state ferries arrive and depart Sitka only when tides are favorable at Sergius Narrows.

All this watery commotion is the result of friction—the rubbing, scraping, bouncing-off process that happens all around us. Friction is why our car tires go bald, why rubbing our hands together creates heat, why the water's surface ruffles when wind blows across it. Physicist Richard Feynman said that friction wears out our leather shoes because "the little notches and bumps on the sidewalk grab pieces and pull them off." He also said that we could look at this rubbing as an exchange of energy. Through friction, our shoes lose energy to the sidewalk.

The same could be said about the tide: through friction, it loses energy—*lots of it*. Some is absorbed in the ocean floor as heat, but most of it exerts a torque on the earth's rotation, slowing ever so slightly the length of our days and, in turn, causing the moon to speed up and spiral away.

Thus, the moon, which *causes* the tide, is in turn *pushed away* by the tide.

What we witness when we sail through narrows is friction at work. Every whirlpool, every eddy, every dimple of tension is evidence of energy moving from the moon to the water and back to the moon.

At my cabin table, awash in moonlight, I browse the pages of the *Nautical Almanac,* where rows of astronomical figures indicate the positions of the sun, moon, and earth. I see that on May 25 the moon will be fifteen days old—full, and at perigee. In the *Tide and Current Tables,* I look up Seholt Narrows, a legendary passage about a hundred miles north of my home in Washington State's San Juan Islands. The Coast Salish people called these narrows Skookumchuck, or "fast water." The tables confirm the celestial event: during ebb tide, water will rush out of Skookumchuck at fourteen knots. During the flood, it will race back in at more than sixteen knots. Skookumchuck's average is half this: eight knots—which is considerable in itself. The almanac's predictions for May 25 are for the largest tides and fastest currents of the year.

This is an extraordinary event I don't want to miss.

～

After Newton, scientists learned that the study of tides could be separated into two parts: what happens *up there* (astronomy) and what happens *down here* (fluid dynamics). A hundred years after Laplace, we learned that we could further separate what happens *down here* into at least two parts: the fluid dynamics of the deep ocean and of the shallow continental shelves.

Common sense tells us the line between land and sea is drawn on the beach, or in the intertidal zone. For most practical purposes, it is. But when we think about global tides, that line shifts to the outer edge of the continental shelf. This edge, often tens of miles offshore and buried under water, is where the continent truly surrenders to the deep ocean. Depths over the shelf are usually less than 100 fathoms (600 feet); beyond the shelf they tumble into the abyss.

Continental shelves vary in width. On the U.S. west coast, they average 20 miles. North of Cape Hatteras on the east coast, they're about 150 miles. Off southern Florida, they're only a few miles. Patagonia, eastern Australia, and the Arctic have wide shelves. At almost a thousand miles, the Siberian Shelf (which includes the Bering Straits) is the world's widest. Sumatra and the west coast of Chile have almost no shelf, nor, as a general rule, do offshore islands, such as Hawaii, Bermuda, Tahiti, the Azores, and the Maldives.

The shelf's edge, then, defines not just the true edge of a continent but the deep ocean basins. This is a critical tidal boundary. What the tide does on one side of this line is different from what it does on the other, a difference ultimately defined by how much or how little the tide feels the bottom or is obstructed by landmasses. In the deep ocean basins, the tide feels less of the ocean bottom and, with few land-masses to bump into, is generally free to race around unobstructed. By the early twentieth century, scientists had confirmed that tide waves do, indeed, whisk about the large ocean basins. It was known that they circle amphidromic systems, which are sprinkled everywhere: in deep water *and* in shallow water. And it was known that these fast-moving waves are not tall. In the deep ocean, the average high tide is less than two feet, worldwide. This is why offshore islands, with no complicat-ing continental shelves, have fairly small and regular tides. Ascension Island's average tide is about two feet, as is that of the Maldives, Samoa, Bermuda, Guam, and the Hawaiian Islands. Tahiti's tide is just over a foot.

But when the fast-moving, low-hump oceanic tide hits the conti-nental shelf, it distorts just like an offshore swell hitting the reef at Mavericks. It slows, steepens, twists, bends, refracts, reflects, and defracts. The shallower the water, the more distorted the tide. Hence, the farther the tide travels inland, the stranger its behavior becomes and the harder it is to predict. A bore moving up a river is a tidal dis-tortion at its most extreme.

There are other, less well-known distortions. As the oceanic tide warps over the continental edge, it forms mountainous waves below the surface that march in slow motion, like wax in a lava lamp, across the ocean, often breaking on distant shelves or ridges. Called *internal tides*, these waves are sometimes 1,500 feet tall, crisscrossing the seas virtually unseen except where they occasionally show on the surface as bands of smooth and ruffled water. "Internal waves, which are sometimes created by deep-keeled ships but most often created by the tide, were discovered in the Norwegian Sea in the early twentieth century," explains Sally Warner, an oceanographer at Oregon State University. "A great deal of research is currently being done to understand them and their impact on global oceans."

Warner, who wrote her doctoral thesis on internal waves generated by tidal currents in Puget Sound, explained over the phone that surface waves like those at Mavericks occur at the interface of water and air. Because water is eight hundred times denser than air, this interface—or *density layer*—is distinct and fairly stable. But there are other density layers below the surface, formed by layers of water with differences in temperature and salinity. These are the layers where internal tide waves are found.

"We know that these waves are born where strong tidal currents flow over steep and rugged bathymetry, like over shelves and ridges, or even around headlands in inland waters," says Warner. "It's hard to track an individual internal wave, but recent studies in places like the South China Sea—where the largest internal waves have been discovered—are shedding light on how they're formed and where they go. They play a role in global mixing—pulling warm water and oxygen down from the surface and pushing cold water and nutrients up— which is incredibly important to the biological health of the oceans."

Some scientists go further, suggesting a cause-and-effect relationship between internal tides, mixing, sea surface temperatures, and global climate. "It's been known for a long time that climate is influenced by sea surface temperatures," said oceanographer Malte Müller. "Since we know that tidal mixing is one of the drivers of sea surface temperature, in the future we may find a correlation between tide cycles and global weather patterns."

～

Without friction, tide theory would have been put to rest a long time ago—perhaps with Newton. England's Southampton wouldn't have a double high tide; Holland's Scheveningen wouldn't have a double low tide; Victoria, British Columbia, wouldn't have once-daily tides on some days and twice-daily tides on others. Chesapeake Bay, too, wouldn't have a high tide at its mouth and head—Norfolk and Baltimore—at the same time, with 120 miles of low tide between.

Friction, in fact, is the reason scientists chew pencils and pull their hair, why Aristotle may have drowned himself, and why contemporary tide-modelers suffer "oh god" moments.

"The whole system is sticky," says oceanographer Chris Garrett. "The further inland you go, the stickier it gets." Scientists often call this stickiness *nonlinear shallow-water processes*, an expression that elicits commiserating nods. "Nonlinear" means it doesn't follow a logical progression: one plus one plus one doesn't always equal three. We may not understand every nuance of the tide's nonlinear behavior, but at least we have a name for it.

Stickiness, or friction, is indeed widespread. Not even the deep ocean is free of it. According to Newton's theory, the two watery tidal humps—one under the moon and one on the opposite side of the earth—would travel a thousand miles per hour to keep up with the moon. They would need an ocean about six thousand miles deep (with, of course, no continents to bump into). In that deep ocean, the tidal humps would stay directly under the moon. Standing on the beach, we would witness high tide at the same time the moon passed overhead. All would make perfect sense.

But at an average depth of three miles, even the deep oceans aren't deep enough to allow this. Consequently, the bulges—the tides—drag on the bottom and slow down, even over the nearly seven-mile-deep Mariana Trench, the deepest part of the ocean. The tides are always behind and always rushing to catch up. When we see the moon overhead, we know high tide is coming soon, but we don't always know *how soon*.

In the first century CE, Pliny described this "lagging" as "the effect of what is going on in the heavens being felt after a short interval, as we observe with respect to lightning, thunder, and thunderbolts." The Chinese recognized the delay too and accounted for it in the world's first known tide chart, which logged the arrival of the Qiantang tidal bore. The chart, published in the first millennium, showed the largest tidal bore arriving a couple of days after spring tides.

Whewell, the nineteenth-century British scientist, called this phenomenon *the age of the tide*. He was referring, as were Pliny and the Chinese, to the delay between full or new moon and the arrival of the largest spring tides. Where I live, on the Pacific Northwest coast, the tide's age is about two days. On the Atlantic coast and in Europe, it's about a day and a half, and in Australia it's about three days. In places like the South Pacific's Solomon Islands, the age is *minus* a day or two, which is the result of a combination of factors, such as the proximity of an amphidromic hub, an underwater canyon, or some other nonlinear shallow-water process.

Worldwide, the tide's *age* varies from plus seven days to minus seven days. Spitsbergen in the North Atlantic, for example, has an age of minus seven days, and parts of Patagonia, with its large continental shelf, have ages of plus seven days.

The tide's *age* refers to the global oceans as a whole, but there are local delays, too, that are often exaggerated as the tide progresses into stickier waters. On the outer Pacific coast, for example, the tide circles a large amphidromic system that sits off Baja, Mexico. The long, spidery arms of that system—each arm representing a high tide—reach toward the coast, making their first landfall at Baja's Bahia Magdalena. From there the tide hastens north over the gently sloping beaches of southern California, pressing toward Cape Mendocino and Cape Flattery at the entrance to the Straits of Juan de Fuca, near where I live. If high tide arrives in San Diego at 8:30 a.m., it will arrive in San Francisco two and a half hours later and slip under the fishing fleet in Sitka, Alaska, about an hour after that. It sweeps the whole

1,500-mile coast in about three and a half hours. As the amphidrome continues spinning, it brushes across the Aleutian Chain and heads toward Russia, Japan, and China before completing a full circle, only to begin again.

When this same tide hits Cape Flattery on the northern coast of Washington State, some of it branches off and elbows into the Straits of Juan de Fuca and other inland passages. It will take at least four hours to traverse the 130 miles to Vancouver, British Columbia, and another hour and a half to reach Egmont and Skookumchuck. There, with the increasing friction, it stalls and bucks and stumbles as it does at hundreds of other pursed-up narrows. It piles up at the entrance to Skookumchuck, creating a water level difference of two or three feet between one end of the sill and the other. This difference, called a *hydraulic head*, forces billions of gallons of seawater to boil through the narrows.

In spite of the apparent rush, the tide takes several hours to squeeze through Skookumchuck. At 7:30 p.m. it finally comes to a rest at the inlet's head. In all, it takes eight hours to cover the two hundred miles from Cape Flattery. It takes less than thirty minutes to cover the same distance on the outer coast.

～

When the time comes to sail north to Skookumchuck, I load *Crusader*'s galley with provisions, top off the water and fuel tanks, lug the charts and reference books from my cabin and stow them in the pilothouse. Now all I have to do is untie the lines and go. Although I usually sail with a crew, I'm taking this trip alone. The solitude will allow me more flexibility and focus.

My plan is to make the hundred-mile trip to Skookumchuck in two days, anchoring overnight in the Canadian Gulf Islands. Once I arrive at the small harbor of Egmont, I'll moor *Crusader* and use her sixteen-foot skiff to venture the last few miles into the narrows where I'll stay to see a full tide cycle, from maximum ebb to maximum flood.

The skiff, much smaller and lighter than *Crusader*, will be easier to maneuver in the narrows. Its ten-horse outboard should be plenty strong to overpower the rapids. I think.

There's another important difference between *Crusader* and the skiff. *Crusader* has a displacement hull, meaning she has a long deep keel (seven feet deep, in her case). This keel, a feature she shares with all sailboats, trawlers, and big ships, is efficient for moving through the water at slower speeds, but it acts like an underwater sail, responding to currents the way canvas responds to wind. In turbulent water like Skookumchuck, it's easy to lose steerage and fishtail out of control.

The skiff, like most get-up-and-go powerboats, has a flat bottom with no keel. It's less stable, but it can skip along the surface without feeling much below. In fact, it responds more to the wind than to subsurface currents. (When sharing an anchorage, flat-bottom boats and those with keels don't always mix well. Sailboats, for example, will circle with the tidal current, while powerboats will circle with the wind. The wind and tide aren't always in sync, so in tight anchorages sailboats and powerboats can tangle. The turn of the tide or a change in wind direction brings skippers topside to find out if they've anchored with enough room to swing—or if their neighbor has.)

After a morning at home with my wife and son, I untie *Crusader*'s mooring lines and motor into Georgia Strait, the large body of water that forms the backbone of the Salish Sea. By noon a stiff southwesterly kicks up. I usually wouldn't raise *Crusader*'s large and heavy sails when single-handing, but I can't resist these conditions. I hoist the sails one by one: the mainsail first, then the foresail, the staysail, the jib, and, finally, the flying topsail. It takes me almost an hour, but with all five sails billowing, *Crusader*—the heavy old halibut schooner that she is—seems to lift with pride. I shut the engine off, and for hours we skate along on a broad reach, her bow knifing the dark water as if slicing through ripe fruit, throwing foam like laughter across the bow.

Crusader's motoring speed is eight knots. Today, under sail, she's doing five. With a flooding current of two knots, she's making seven across the bottom. If the tide were ebbing at the same speed, we'd be making only three knots over the bottom. That's a big difference. Today, with the wind and tide in my favor, I'll cover about forty-five miles. It would be less than half that if I were sailing against an ebb.

Before the sun sets, I round up into the wind and wrestle the sails down, again one at a time—flying topsail first, mainsail last. I do this in the open water of Georgia Strait, where I can drift without worrying about hitting a rock.

When I finally drop anchor inside a protected bight, I know that I've been given a perfect day of sailing—something that rarely happens on this wind-fickle coast. I sit on *Crusader*'s deck and watch other boats plying north. Most of them, I imagine, are embarking on the 600-mile migration to Alaska, their hulls freshly corked and painted for the long commercial salmon season ahead. Hundreds of boats of all sizes and shapes make this yearly journey. Although a handful head north for pleasure, most are part of the fishing fleet—seiners, trollers, packers, gillnetters—that leave throughout the spring and return, scarred but laden with dollars, in September.

For eleven seasons I joined them. I remember well the fresh sense of adventure when I got underway, coupled with a slight apprehension of what the season would bring—knowing we would peel away layers of tameness with each mile left on our stern; knowing we would encounter rough seas, torn sails, failed gear, engine trouble; knowing too that our lives would be deepened by beauty, wildness, and solitude beyond measure.

〜

The next morning I crank *Crusader*'s 150-pound anchor aboard and motor toward the snow-topped peaks of the coastal range and the narrowing channels of Sechelt Inlet. The tide is flooding at about three knots when I enter Agamemnon Channel, but you'd never know it by looking at the surface. In this wide, 600-foot-deep passage, there's no sign of friction: no boils or whirlpools, no backeddies. When I turn the last corner into Egmont, where I plan to moor *Crusader*, I see boils erupting in the channel's center and I feel *Crusader* shifting subtly from side to side, the way an airplane moves in turbulent air. Skookumchuck is a couple of miles away.

In Egmont, I tie off at the government wharf. When I walk into the wharfinger's office to pay overnight moorage, Vera Grafton is sitting at her desk in a loose T-shirt and turquoise sweatpants, eating cherries and swatting flies. A seventy-five-year-old Nuuchahnulth Indian who has lived in Egmont since she was eight, Vera wears her long gray hair draped loosely around her face. She spits a few pits into a plastic bowl, then offers me a chair and, eventually, some cherries.

She laughs when I tell her I'm planning to take my skiff into the narrows the next morning before light. "I see lots of people—kayakers, rafters, sailors, even hikers—come to see the rapids," she says. "I understand why you're drawn, but you're crazy to go out there alone." I sense her eagerness to share tales of caution, but I'm relieved that at least for the moment she chooses a more poetic story.

"According to the Salish Indians," she says, leaning back in her chair, "a beautiful young girl, Ko-Kwal-alwoot, follows the tide's whisper into the churning waters of Deception Pass, a narrow channel down south. She enters slowly, first to her ankles. She's afraid of being swept away by the current, but when she's waist-deep, a hand from a guiding spirit reaches up. It offers itself and gently draws her in." After a number of meetings, the young girl falls in love with the spirit, Vera tells me, and accompanies it into the undersea world. Today, the Samish tribes believe Ko-Kwal-alwoot and her spirit-lover look after the people's welfare. The turbulence in the water is Ko-Kwal-alwoot's long hair, reminding the people of her presence.

Vera gazes out her small window and spits a few more pits in her bowl. "Like Ko-Kwal-alwoot, I think many of us are drawn to these waters. There's a spirit in them, a life, something that whispers to us. They're dangerous, but maybe that's part of the attraction. I don't know."

Out Vera's window is a scene that could be found almost anywhere on this coast: mossy boulders, thick cedar and hemlock forests, a fleet of rusted trawlers in the harbor, frigid blue-black water fingering the channel's edges. I can't hear it, but I know how it sounds: the water's lulling gurgle, the light snapping of halyards against masts, a bald eagle's little-girl-like cry. If I could smell it, there would be tree resin warming in the sun, salt, iodine, and fish. As familiar as the scene is, it offers no clue of what's happening around the corner at Skookumchuck.

"I've seen a lot of stuff go wrong out there," Vera says, breaking the silence. "Just last year a tug pulling a heavy scow tried to go in at the wrong tide. They lost control and flipped in seconds. Three of the four crew got out right away, but one was trapped under the hull for a few minutes. The engines were revving and props turning. Smoke was pouring out of the cabin. Everyone got out OK, but they were real shaken up. No one can survive for more than a few minutes in that cold water."

"I remember an accident back in the fifties," she says as she slides the bowl of cherries toward me. "A teenager named Ruth and her three- and four-year-old sisters, Jane and Patsie, rowed from their home a few miles up the inlet to get ice cream for the youngest's birthday. They never made it home. A few days later, the ice cream bucket and two pinafores surfaced, circling in the tides."

~

When Vera and I finish the last cherries, she hands me the moorage receipt and wishes me a safe trip. That evening, as I prepare *Crusader*'s skiff for the morning's adventure, I think of other tide accidents. One of the earliest accounts on this coast is from the French explorer Jean de la Pérouse, who in 1786 sailed his two ships, the *Boussole* and the *Astrolabe,* into southeast Alaska's Lituya Bay. He was warned of the area's dangers by the local Tlingit people who, like dozens of other Native tribes, were highly skilled at navigating the tides. The Tlingit told Pérouse that large creatures shook the waters at the bay's mouth until they became impassable.

"These Indians seem to have considerable dread of the passage," wrote Pérouse, "and never ventured to approach it, unless at the slack water of flood or ebb." Even then, a shaman "would stand and raise his hands to the heavens while the paddlers pulled like the devil to get through."

Pérouse sent three small boats to "sound the passage . . . and measure its width" but admonished his crew not to approach the passage if "there were the least appearance of breakers, or even swell." But the boats got too close to the opening during a strong ebb and were sucked into a "fury of waves." One boat survived, but the other two, along with twenty-one crewmen, were lost. No remains were found.

Pérouse had sailed around the world without losing a single man. After the Lituya Bay incident he wrote: "Nothing remained for us but to quit with speed a country that has proved so fatal."

Six years later, Captain George Vancouver sailed through Seymour Narrows, one of the coast's most infamous rapids, near present-day Campbell River on Vancouver Island's east coast. Peak currents at Seymour exceed fourteen knots. Miraculously, Captain Vancouver must have sailed through at slack water, since he reported the channel safe for navigation. He missed by a matter of hours the treacherous conditions that caused so much grief for future sailors. He also missed Seymour's most menacing feature: Ripple Rock, which at low tide rose in the channel's center to within nine feet of the surface and caused increased turbulence and a grounding hazard for deep-draft boats. One of the rock's early victims was the U.S. Navy steamer *Saranac*, which sailed into Seymour Narrows in 1875. A crewmember wrote, "Here the contending currents take a vessel by the nose and swing her from port to starboard and from starboard to port as a terrier shakes a rat." Out of control, *Saranac* struck Ripple Rock and sank. The rock eventually destroyed or severely damaged more than twenty-five large ships and countless smaller vessels, at a cost of 114 lives.

By the twentieth century it was decided that Ripple Rock had to go. Engineers agreed the best way to accomplish this was to drill into the rock and set explosives. But how? It was tried from above, but no anchoring or cabling system was strong enough to hold a barge in place in such turbulent conditions. After twenty-five years of studies of such failures, it seemed the only workable approach was to drill horizontally from shore and come up under the rock. It took three years to do this. Charges were finally set, and a detonation time of 9:31 a.m., April 5, 1958, was chosen to coincide with a perigean spring tide's extreme low water and strong ebb. The low tide (less water over the rock) would allow for a larger explosion and a larger dispersal of debris. A strong ebb would flush the debris northward, away from Campbell River.

It worked. And it should have. It was the largest peacetime non-nuclear explosion to date.

Although Ripple Rock was gone, the tide wasn't. Accidents at Seymour continued to accrue, as they did at dozens of other coastal rapids. Today the Canadian Rescue Center fields more than three hundred emergency calls a year, many coming from vessels caught in places like Seymour.

What happens at Seymour happens worldwide. Treacherous passages include Norway's Saltstraumen (up to twenty-eight knots) and Moskstraumen (twenty knots), Japan's Naruto Strait (twenty knots), Nova Scotia's Minas Basin (sixteen knots), Scotland's Pentland Firth (sixteen knots), and South America's Strait of Magellan (fourteen knots). Today the *North Sea Pilot* gives the very same warning for the Pentland Firth as it did in 1875:

> Before entering the Pentland Firth all vessels should be prepared to batten down, and the hatches of small vessels ought to be secured even in the finest weather, as it is difficult to see what may be going on in the distance, and the transition from smooth water to a broken sea is so sudden that no time is given for making arrangements.

These places have loomed in sailors' minds for centuries, and the names show it: the Chopper, the Man-Eater, the Swallower, the Old Hag, the Toilet Bowl, the Bitch.

〜

I want to see Skookumchuck at first light, so I wake at 3:30 a.m. At this hour Egmont is quiet and dark, save a yellowish streetlamp that hovers over Bathgate's General Store. A string of Christmas lights droops from the eaves, its bulbs mostly blown or missing. Vera won't open her office for another few hours. I throw a life jacket, sandwich, and bottle of water into the skiff and untie the lines. Not wanting to wake anyone in the marina, I row out to the channel before starting the outboard and ride a flooding tide toward the narrows. My plan is to stay out all day, long enough to see Skookumchuck at maximum flood and ebb.

A few minutes after I get underway, a green navigation light appears off the bow. It flashes every four seconds, letting me know that I am steering directly for one of the islets in the middle of the Narrows. Beyond it is the quick flashing light on Skookum Island, two and a half miles farther up the inlet. A purplish line marks the 6,000-foot ridges of Earle Range in the west. I can see no details below that and nothing at water level.

Suddenly the four-second green light, which must be less than a quarter-mile away, disappears. Spooked, I shut down the motor and drift, hoping to regain my bearings. I have no idea what has happened until my eye catches the silhouette of a small islet directly in front of me. While I thought I was making headway toward the green flashing light, the skiff was caught in a cross-channel current, pulling me off course.

The water is calm in the lee of the islet, where I pause for a moment to observe what's happening around me. The narrowest part of Skookumchuck is close enough that I can hear the rising tide's hiss. I remember from the chart how it wraps around the headland and skirts the midchannel island—the one locals call Tremble Island—before exploding into a frothing vortex. I don't want to get sucked into this, at least while it's still dark. If I can nose into the lee of Tremble Island, I'll find calm water and a place to tie up my skiff until daybreak.

As I maneuver from the islet, the green flashing light reappears. I speed up to cross the current, and just as my bow nudges the rocky edges of Tremble Island a whirl of stars comes at me from below, like dust in a light wind, engulfing my skiff in bioluminescence. The boat twirls, leaving a prop-wash trail of light, but I'm able to weave my way to shore.

I tie the skiff to some rocks and explore the small island. On my tide chart, I confirm that the flooding tide is about to go slack. It will start ebbing shortly after daylight, go slack six hours later (around noon), and then flood again all afternoon.

When the sun rises, I motor up the inlet against a mild ebb current and cut the engine at Highland Point. From here, I'll drift the three miles back to Skookumchuck Narrows, arriving when the current peaks. For now, the scene is peaceful and unthreatening. Milky water from inland glaciers rounds this point and meets the last of the incoming oceanic tide, creating a distinct tide line, stippled with capillary waves and bits of drift. Phalaropes and seals feed along its nutrient-rich edge.

My skiff drifts silently in three or four knots of current. Except for the mainland trees silently passing by, I have no sensation of movement. The boat moves with the water, gliding fluidly over the bottom 1,200 feet below. But I know everything is about to change. In less than a mile, the water's depth vaults to thirty feet over Skookumchuck's sill, then drops again into the depths of Agamemnon Channel. Billions of gallons of water, pushing to get out, will funnel to the channel's center and warp over the sill like a giant tongue before collapsing into a sunken disk downstream. Although I'm a half-mile upstream of all this, I start the outboard and let it idle. Just in case.

≈

I think again of my conversation with Vera Grafton, especially the story of Ko-Kwal-alwoot. It's a myth, of course, but it speaks to a longing, a fascination that I too feel. Of all the miles I've logged on this coast—all the anchorages, circumnavigations, and open-water crossings—what I remember most are the pinched-up tidal passages. They become focal points, not just of moon and tide and boat, but of human memory and experience, hourglasses of the imagination, small openings where life is concentrated.

Some of history's most prominent literary figures—Homer, Edgar Allan Poe, George Orwell, Victor Hugo, Jules Verne, Sir Walter Scott, Italo Calvino—had encounters with tidal rapids, some imaginary and some real. After five years of wandering the Mediterranean Sea, Homer's hero, Odysseus, had to sail through the boiling waters of Charybdis to make it home to Ithaca. There was no other way. Charybdis was hatched from Homer's imagination, but it was likely inspired by the raging currents in the Strait of Messina at the toe of Italy's boot.

In 1947 George Orwell almost drowned, along with his nieces and nephews, when he took a small, open boat into Norway's Moskstraumen. The violent sea flipped the boat. Luckily, they were next to shore. Orwell surfaced with his three-year-old nephew and they all scrambled to safety.

Poe's imagination was ignited by this same passage. In 1841 he wrote *A Descent into the Maelstrom*, a gripping tale about an old man whose ship was sucked down a whirlpool at Moskstraumen. The old man lashed himself to a barrel, which floated to the surface. The ship and crew were lost, but he survived. Later, looking down on the passage from a cliff, the old man tells his companion, "The six hours of deadly terror which I endured have broken me up body and soul. You suppose me a very old man—but I am not. It took less than a single day to change these hairs from a jetty black to white, to weaken my limbs, and to unstring my nerves."

I haven't been sucked down a whirlpool, but I've had my share of

These color photographs dramatically illustrate tidal range across the United Kingdom.
Each pair is taken from the exact same position—at low tide (top) and high tide (bottom).

PERRANPORTH, CORNWALL

August 29 and 30, 2007. Low water 12 noon, high water 8:00 p.m

LYNMOUTH, DEVON

September 17 and 19, 2005. Low water 12:45 p.m., high water 7:30 p.m.

CROSBY, LIVERPOOL

April 5 and 6, 2008. Low water 9:00 a.m., high water 12 noon.

SALMON FISHERY, SOLWAY FIRTH, SCOTLAND

March 27 and 28, 2006. Low water 5:20 p.m., high water 12 noon.

ST MARY'S LIGHTHOUSE, WHITLEY BAY, NORTHUMBERLAND

September 17 and 20, 2008. Low water 1:00 p.m., high water 5:50 p.m.

STAITHES, YORKSHIRE

September 14, 2004. Low water 9:45 a.m., high water 4:30 p.m.

CUCKMERE HAVEN, SUSSEX

August 12, 2006. Low water 9:15 a.m., high water 2:50 p.m.

TIDAL ROAD TO SUNDERLAND POINT, LANCASHIRE

March 29 and 30, 2010. Low water 6:15 p.m., high water 1:00 p.m.

humbling experiences in the tide. At Nakwakto—perhaps the fastest tidal passage on the coast—I once shut *Crusader*'s engine off in the middle of a peak ebb, just to listen and experience the churning, twisting power. Grabbed by the current, *Crusader*—her sixty-five feet of oak and fir and rigging—listed, groaned under the strain, and spun downstream.

One summer when my mother and father joined me, I motored *Crusader* into Yaculta on the mid–British Columbia coast at the wrong tide. I was less experienced then, and instead of turning back I continued. It was a bad decision. I couldn't make headway against the current. We lost steerage. My father and I gripped each other's hands as we came about in the middle of the treacherous seaway. Fortunately, we got out without incident and waited in a nearby anchorage for the next tide change.

We never were very close, my dad and I. Neither of us seemed able to accept the other for what we were—he wanted more of me and I wanted more of him. The experience in Yaculta seemed to cut through all that, at least for a few moments.

≈

As my skiff approaches Skookumchuck, I put the outboard in gear and find a backeddy where I can pause my progress on the ebb and observe from a safe distance. The main current in the middle of the channel, which is a quarter-mile wide here, is racing at ten or twelve knots when it hits the first of several islets in the narrows head on. Some of it is diverted down a chute and breaks into rapids over Canoe Pass. The rest curls over slick boulders and collides with a ridge of faster-moving water in the channel's center. The colliding currents chafe, tearing off a train of eddies. Most of the fast-moving ridge slams into Tremble Island, which lies broadside in the narrowest, shallowest part of the passage.

Driven from hundreds of feet below, cold water rushes up and boils against the shore. All that momentum has nowhere to go, so the

Skookumchuck rapids at peak flood.

water folds back on itself, its surface taut as a drum skin. Some joins the outflow of Canoe Pass as it drains into Agamemnon Channel. But most is sucked sideways, through bull kelp, plowing into green-water whirlpools. Fed with power, these whirlpools, initially several feet in diameter, dilate instantly to twenty times that size. They dominate the channel for minutes at a time. Inside each whirlpool, water speeds up, sinking four or five feet as it corkscrews toward the pool's center. It vanishes into a dark hole the size of a basketball. Whirlpools occasionally get hold of bull kelp and twist it so tightly that it remains bound even after the whirlpool has dissolved. Beached at high slack water, these serpentine spirals are mute reminders of the tide's ferocity.

Downcurrent, a backeddy sweeps foam and Bonaparte's gulls nonchalantly alongshore. I hover near shore too, venturing out only to ride the raised edge of a whirlpool or nose into its dished-out center, hearing the old man in Poe's tale: "The edge of the whirl was represented by a broad belt of gleaming spray; but no particle of this slipped into the mouth of the terrific funnel, whose interior, as far as the eye could fathom it, was a smooth, shining, and jet-black wall of water . . . sending forth to the winds an appalling voice, half shriek, half roar."

I've never heard of a whirlpool swallowing a boat in one gulp. Logs and seaweed, yes, but not whole boats. My fear is not that; it's getting swamped by a wave or thrown overboard. If I fell into this icy cauldron, I'd have about fifteen minutes before growing weak and confused with hypothermia. I'd have to get out before then.

≈

Skookumchuck's dramatic tides hint at a much larger, fantastically complex process that ripples through the solar system. The effect of global tidal friction actually acts as a brake on the earth's rotation, making each day 1/50,000 second longer. And as *tidal braking* slows the earth's rotation, energy is transferred by angular momentum to the moon, speeding it up and pushing it away at a rate of about an inch and a half per year, or ten feet in a human lifetime.

These numbers may seem negligibly small. And measured day to day they are. But over geological time, they add up. Scientists believe that 400 million years ago the earth's day was only twenty-one hours long and the moon was about ten thousand miles closer. To earth's earliest inhabitants, the moon would have loomed 20 percent larger than today. Tides, too, would have been larger and more frequent.

Richard Brathwaite didn't know that time was slowing down when he wrote "Time and tide stayeth for no man." In his day—the early seventeenth century—it was commonly understood that time and tide were reliably constant. Neither could be changed or stopped. The twenty-four-hour day was the time it took the earth to make one full rotation, the month was the time it took the moon to orbit the earth, and the year was the time it took the earth to circle the sun. By the late seventeenth century, Isaac Newton had described a well-behaved universe whereby time and motion followed predictable rules. And while

Newton would have understood the concept of friction elsewhere, it was not factored into his tide model.

But less than a decade after the 1687 publication of Newton's *Principia Mathematica*, Edmund Halley noticed that something was awry in the heavens. An accomplished scholar who served for twenty-two years as Britain's astronomer royal, Halley discovered that ancient eclipses had occurred earlier than existing calculations indicated they should have. If the earth rotated at a uniform speed over geological history, Halley reasoned, it would follow that eclipses, which are the result of highly specific and predictable alignments of the sun and moon, could be traced to an exact place and time. But when he did the math, it turned out not to be so. Was time changing?

Luckily for Halley, eclipses had been observed and recorded for at least several thousand years. The Babylonians, Chinese, and Arabs kept the earliest surviving records, many of them etched by priests on clay tablets or parchment. In these, Halley found discrepancies. An eclipse recorded in Syria in 910 CE, for example, occurred half an hour earlier than it should have.

Perplexed, and under the influence of his friend Newton's orderly theories, Halley pursued the subject only so far as to request eclipse observations from other travelers. It never occurred to him that his discovery was evidence that the earth's rotation *was slowing down* and the length of a day was growing longer, much less that tides were the cause. If this were true—as we now know it is—then *time itself* could no longer be seen as constant and predictable.

Progress in understanding Halley's discovery came slowly and in pieces over the following 250 years. For most normal people, including those of us living today, the concept of variable time is hard to digest. Physicists and astronomers are better equipped to understand it, but the rest of us don't think of time as dependent on how much friction there is in the daily movement of tides across our watery planet. Yet it is.

In 1754, a half-century after Halley, the Berlin Royal Academy of

Sciences put forth the following questions: "Has the earth undergone any alteration since the first period of origin? If it has, what would be the cause, and what can make us certain of it?" A response came from a twenty-four-year-old philosopher of metaphysics, Immanuel Kant, who suggested that the earth's rotation had indeed fluctuated over time. Tides, he predicted, were the cause.

Another century passed before the thread was picked up again, this time by physicist and astronomer George Darwin (Charles Darwin's third son), who was a scholar of lunar history. In his 1889 book *The Tides and Kindred Phenomena,* he hypothesized that roughly a billion years ago the earth had no moon. The day was only a few hours long, and hundred-foot tides sloshed so violently that they ripped the moon from the bottom of the Pacific and heaved it into space, only to be captured by the earth's gravity while still close—a quarter the distance it is today. According to Darwin, the earth's day was lengthening and the moon drifting away ever since, due to tidal friction and, to a lesser degree, friction in the body of the earth.

Darwin's theory of the moon's origin, known as *fission,* was disproved in the early twentieth century, but his hypothesis that tides were the cause of the changing orbital speeds and distances of the earth and moon is today accepted fact. Darwin understood this, but he couldn't prove it.

Proof—or the hope of it—came a half-century later when Cornell paleontologist John Wells discovered a curious ring pattern in reef corals. "Corals," he wrote in 1963, "lay down daily rings made of calcium carbonate much like a tree lays down rings during the fast- and slow-growing seasons of the year. Some coral species lay down both daily and yearly rings." Thus, just as we can count tree rings to determine their age, we can count certain coral rings to determine the days per year.

Coral, composed of colonies of small animals, have lived on earth for a very long time, perhaps more than 500 million years. Each generation builds atop the one that just died, so over time a mountain is formed, with the youngest coral on top and the oldest at the bottom. The youngest—the ones living today—have ring counts of between 360 and 375, which roughly corresponds with the number of days in our current year. Coral that lived 600 million years ago, Wells found, have rings corresponding to a twenty-one-hour day.

Wells's findings, admittedly rough, appeared in the journal *Nature* in 1963, inspiring a search for long-term records in other living sea fossils, such as scallops, mussels, clams, and algae.

Meanwhile, another source of proof appeared, this one from a reflector left on the moon during the 1969 Apollo landing. The device, not much larger than one of Buzz Aldrin's nearby footprints, was designed to reflect a laser beam transmitted from earth. The beam's travel time—almost the speed of light—could be converted to distance with accuracy within an inch. Over the last forty-seven years, the data has confirmed that the moon is receding at a rate of an inch and a half per year. It was a tremendous step forward in the effort to understand the earth-moon history, but forty-seven years is very short in geological time—not even a blink. What happened in the 4.5 billion years before that? Has the moon's retreat been consistent since the beginning?

These questions are still largely unanswered, but research inspired by Wells continues. In the 1970s physicist Stephen Pompeii and geophysicist Peter Kahn thought they had found an animal that fit the criteria of being both very old and tuned to cosmic rhythms. The chambered nautilus, a relative of the squid and octopus, lives deep in the sea by day and swims to the surface by night. Not much is known about these creatures except that they've been around for about 500 million years.

A cutaway view of a *Nautilus pompilius* showing its spiral structure and growth chambers.

"As nautilus grows," wrote Pompeii and Kahn, "it incorporates two kinds of rhythmically repeating structures. One is the enlargement of the body chamber . . . and the other is septation, by which the parts of the shell abandoned by the animal are partitioned off into chambers." Chambers are formed monthly, each sealed off with a wall as the animal spirals outward. Daily growth rings are laid down within each chamber. When Kahn and Pompeii counted these lines in nautiloids ranging in age from 25 to 420 million years, they found general agreement with Wells's work.

Evidence of the earth's rotational history has also come from the tide itself. Geologists have long known about tidal rhythmites, which are layers of soft stone formed on the seabed of estuaries or tidal deltas millions of years ago. These were once highly protected areas that allowed the tide to come and go with little disturbance from wind or waves. Consequently, each tide left a layer of sediment that varied in thickness, depending on the tide's size and the current's speed. Spring tides with faster currents, for example, left thin layers, while neap tides with slower currents left thick layers. These layers can be read in a laboratory to determine length of day, variations in tide range, and, by extrapolation, the moon's distance.

In the 1990s very old rhythmites were discovered in Australia, France, the Netherlands, and even Utah, Alabama, and Indiana. A 620-million-year-old Australian rhythmite, when sliced in the lab, showed a day length of twenty-two hours, which correlated to an earth-moon distance of twelve thousand miles less than today's.

Some scientists are critical of this kind of research, particularly the coral and nautilus studies, arguing that ring counts are not reliably attributable to days, months, or years, or that data are biased toward desired results. But even if there's not complete consistency in the ring counts, there's *enough* consistency to make reasonable inferences. So far, it's our only view into the ancient rotational history of the earth and moon.

What rhythmites, corals, nautiluses, and even newly

discovered eclipse records tell us is that, while the day is surely slowing and the moon is indeed drifting away, neither is doing so at a constant pace. Numerous factors account for this. Over geological time, tidal friction varies as continents drift or as periods of glaciation come and go (exposing more or less shallow water). Short-period events—earthquakes, volcanic eruptions, high winds over mountaintops (more friction)—can also jostle the earth's rotation.

It seems our solar system is so tightly interwoven that not even the smallest dip or yaw goes unnoticed. In the end, what we've learned since Halley is that time is not constant or reliable. The earth's day never was and never will be exactly twenty-four hours.

And tides are the reason.

～

At maximum ebb Skookumchuck is boiling. I'm worried that my skiff will be overpowered, so I tie off to a thick strand of bull kelp near shore and wait. A couple of hours later, at the last of the ebb, the torrent subsides. I poke back into the narrows, knowing the tide is about to go slack. The engine is off, and so is my life jacket. It's a relief to be out here without worry. Soon enough, the tide will reverse and the narrows will again come to a boil, this time in the flood cycle. For now, I take advantage of the calm and let the skiff circle aimlessly. I eat my sandwich and watch the changing surface. It's late afternoon and I haven't seen another boat all day.

When slack comes, it blankets the narrows in a magical stillness. It's as if a window, once closed and shuttered, has been flung open and another world lets itself be seen. Tension dissolves. Short breaths become long. A fish jumps; two cormorants take flight, stretching their slender black bodies across the narrows.

If my first encounter with Skookumchuck had been at slack water, whether planned or not, I would probably have slipped through without taking notice, just as Captain Vancouver did at Seymour Narrows many years ago. I'd have no reason to suspect the fury that erupts on this placid water. That's the irony of these places. As sailors, we plan our passages precisely to avoid seeing them in their most menacing moments.

Years ago, I took a boatload of people on a seminar through Alaska's Sergius Narrows on *Crusader*. I talked about the passage for days. We planned our trip around it.

It was raining when we got there. Everyone stood on deck, hooded in raingear, because they didn't want to miss the narrows, which had acquired mythic proportions. One of the foredeck crew came to the bridge to ask when we would arrive at the fabled spot, where the water would surely be raging like Poe's maelstrom.

"We passed it five minutes ago," I answered.

≈

Sailors aren't the only ones who want to get to these places right at slack water. Experienced scuba divers will do the same, often clad in clunky gear and balancing tenuously on a boat's rail, waiting for just the right moment to dive in. For them, the window of opportunity lasts only a few minutes. What's below, however, is worth the trouble.

"It's unbelievably beautiful down there," says Peter Naylor, a biological researcher who has dived at Deception Pass, between Washington's Fidalgo and Whidbey Islands, more than eighty times. "Every inch of the floor is covered with invertebrates: pink and purple sponges, hydro corals, northern feather duster worms, acorn barnacles. I might find one or two giant white anemones on a normal dive, but in Deception Pass I see crowds of them. At Nakwakto up the coast, you can see hordes of red-lipped barnacles, something you can find only in fast-moving currents. And the colors—reds, whites, pinks, purples! It's a fantastically rich habitat, with lots of food zipping by in the current and creatures hanging on for dear life."

As enchanting as it is, the world below is not without challenges.

Accidents related to scuba diving in tidal passages are common all around the world. Naylor, who is a member of the Sheriff's Dive and Rescue Team, says the diving window at Deception Pass, where the currents run up to nine knots, is about twenty minutes. "The currents never really stop," he says. "They slow to about a knot, swirl around a bit, and then suddenly run at one knot the other way. I dive in the space between one knot and one knot; beyond that, it's a hurricane down there. I stayed too long once, trying to pick up a fishing lure. Within seconds I was torn away, turned in somersaults, and slammed against a rock face. Luckily, I got out unhurt."

Not long ago, Naylor was called to find the body of a woman who was lost while diving near Deception Pass. "I really wanted to find her," Naylor says, "because I know how much it means to the survivors. But she was gone, like a leaf in the wind."

〰

When the tide turns and starts flooding at Skookumchuck, I motor up the inlet to Highland Point, where I was earlier and where I hope to see the moon rise between the mountain peaks. Skookumchuck's current will be peaking at sixteen knots in about an hour. Of the year's 706 tides, this will be the largest and fastest.

For now, I relax. The water is fairly calm at Highland Point, so I drift without worry. I finish the last bites of my sandwich and lean back against a thwart. When the moon appears, it seems to pause amid the treetops, as if readying for the long solo passage across the night sky. As big as it is (in perigee), I wonder what it looked like in ancient times, when its distance was a quarter—or even half—what it is today. According to scientists, the earth was raw and steamy and dangerous back then. The moon looked as big or bigger than the sun, and 100-foot tides raced around the planet like the tidal bore on the Qiantang.

The writer Italo Calvino imagined this world in his 1965 story "The Distance of the Moon":

Depiction of Calvino's story "Distance of the Moon."

We had her on top of us all the time, that enormous Moon: when she was full . . . it looked as if she were going to crush us; when she was new, she rolled around the sky like a black umbrella blown by the wind; and when she was waxing, she came forward with her horns so low she seemed about to stick into the peak of a promontory and get caught there. . . .

There were nights when the Moon was full and very, very low, and the tide was so high that the Moon missed a ducking in the sea by a hair's-breadth; well, let's say a few yards anyway. Climb up on the Moon? Of course we did. All you had to do was row out to it in the boat and, when you were underneath, prop a ladder against her and scramble up.

Lulled by tonight's full moon and Calvino's vision of climbing onto her with a ladder (if only I had brought one), I hardly notice that the current has picked up. Skookumchuck, a quarter-mile away, is tugging at my little skiff, spinning it one way and then another, like a piece of drift. Trees race by. I put my life jacket on and start the engine.

When I bring the bow around to face the narrows, everything has changed. In the day's fading light, I can see a mass of whitewater ahead, shrouded in mist. Tremble Island's green light flashes dimly in the distance. Skookumchuck looked so different six hours ago that it might as well have been another place. At maximum ebb, the turbulence is concentrated off Tremble Island. Now, during the flood, water bulges around this point but doesn't break into white water until it gets farther downstream. Even then, it doesn't erupt into the dangerous vortices that overwhelm the Narrows at ebbing tide. Instead, it presses southward, forming a series of six- or seven-foot standing waves off Roland Point. The first few of these remain unbroken, stretching across the channel like mountains of etched black glass. Farther downstream, the waves break and clamor amid sputtering foam.

With less than an hour of twilight left, I approach the narrows and search for a way through. I don't want to be out here in the dark again. As I motor against the current, I fear my outboard will be overpowered. To get a closer look before venturing farther, I follow a back eddy along the shore and tuck behind an islet, the skiff bouncing precariously in the surge. I consider the possible routes: I could cross the main channel diagonally, back and forth, like one might tack a sailboat against the wind. With each "tack," I might make enough headway to eventually get through. With only a ten-horse outboard, however, the risk of capsizing is high.

Another option is to portage across the islet to where the water isn't moving as fast. However, even if I carried the outboard separately, dragging a sixteen-foot skiff over a rocky island wouldn't be easy.

The safest thing to do is to wait for the current to subside. But then I realize there might be one more option, one that tantalizes me.

I circle the skiff to study the midchannel current, which arches like a large tongue over Skookumchuck's sill. This is where the current is moving fastest, but it's also less turbulent than the whitewater at its edges. If I can get out there to the midchannel without capsizing or being swept downstream, and if I can point the bow directly into the current and hold it there, I might have a chance. The question would then be: does my outboard have enough power to make headway against the tide? If it doesn't, I'll have to come about like my father and I did in Yaculta. I'd be catapulted downstream, where I'd have to wait for the next tide change.

I can't resist trying. I back up as far as I can and, at full throttle, jump the skiff over the tongue's edge and land with the bow facing the full force of the tide. Immediately the skiff slips backward as if losing power, taking a wave over the stern. The water's added weight could mean the difference between making it or not, so I bail frantically with one hand as I keep the other on the throttle.

The water is alive. Boils spill upward, bulging and pulsing, tearing at the surface and throwing spray three or four feet in the air. One of these boils lifts the boat unexpectedly and sets it down a few feet to starboard, then shoves it ahead the same amount. Seconds later, a whirlpool sucks the skiff down to the gunwales, as if caught in a vacuum. I gain a little, then fall back, always gauging my progress against the trees on shore. Am I passing them, or are they passing me?

After bailing the boat dry, I kick the gas tank toward the bow and lean forward to pick up a little more speed. I pass a tree, slowly. Then another. One tree at a time, I finally make it to the lee shore of Tremble Island.

Relieved and exhausted, I hike to the island's crest and sit, letting my fingers sink into the warm moss. The moon is well up and the narrows are glistening in the light. As I watch, mesmerized by Skookumchuck's power and beauty, I think of how scientists often talk about the tide's "cascading energy" from the moon to the deep ocean, to the shallow continental shelves, to the wild turbulence of hourglass narrows, and finally to the smallest whirlpools and eddies that curl along shore. We can think of it in reverse, too: every dimple of tension hints at something bigger. In the smallest eddy, we find the moon.

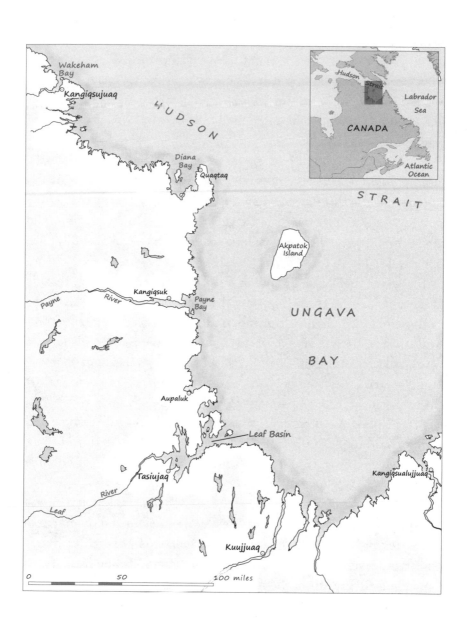

7 Big Tides and Resonance

Fundy and Ungava

Not only is the universe stranger than we
think, it is stranger than we can think.

—*Werner Heisenberg*

It's well below zero and light snow falls as I watch Lukasi Nappaaluk prepare his snowmobile for hunting. He harnesses a plywood sled, called a *kaliga*, to his machine and stows a shovel, rifle, wooden ladder, can of gas, and a *tourq*, a long steel rod sharpened on one end for chopping ice. When the gear is aboard, I swing my leg over the stern of the snowmobile and settle onto the passenger seat. We are ready to leave his village, Kangiqsujuaq, one of fourteen Inuit settlements that cling to the coast of Ungava Bay in northern Quebec, Canada, two hundred miles south of the Arctic Circle. No roads connect the villages to one another, nor to the cities of Quebec and Montreal, a thousand miles south.

Within seconds we drop onto the frozen surface of Wakeham Bay, a two-and-a-half-mile-wide slab of ice that heaves up and down with the tide, leaving its edges strewn with automobile-size ice blocks. We follow a well-worn snowmobile track, which quickly splits in three directions. On late winter days like this, Lukasi and other hunters from his village might follow the track leading inland, hunting for arctic hare and ptarmigan; they might follow the track leading west looking for seal and walrus; or, during the extreme low tides of new and full moon, they might follow the track we have chosen, which leads north four or five miles toward the bay's opening into Hudson Strait. If conditions are just right, hunters on this track will use their *tourq* to chop a hole in the three-foot-thick ice and, for a few hours while the tide is out, shimmy into the hollow regions below the frozen bay to forage for blue mussels.

Lukasi drives the snowmobile cautiously over Wakeham Bay, but the windchill—minus fifteen degrees Fahrenheit—numbs my face and extremities. I'm almost completely covered in layers of wool, fleece, goosedown, and heavily insulated boots and gloves. Only a small wedge of skin around my goggles is exposed, which burns in the cold. We skirt the edge of the bay, slowing several times as Lukasi studies the piled-up ice and snow marking the intertidal zone. On this side of the bay, the intertidal zone is a wide, gradually sloping transition between land and sea. In summer, low tides expose miles of seaweed-covered boulders that provide excellent shellfish gathering. In winter, thick layers of ice accumulate, anchored to the shore. Twice a day the incoming tide forces the ice to float, creating a very active edge. And twice a day the outgoing tide leaves it grounded in a jumbled mess. During the lowest tides, cavities can be found underneath the ice, allowing hunters access into the bountiful intertidal zone below. In midwinter, when summer stores are low and wild game is scarce aboveground, fresh protein is a welcome treat. Sixty years ago, before Europeans moved into the area, hunting under the ice was not just a treat but a matter of winter survival.

Lukasi stops and turns off the engine, asking me to wait while he pokes around. "No," he says, returning, "not right." Several minutes later we stop again. This time he comes back with a hint of excitement on his usually stolid face. We gather the shovel, ladder, and *tourq* and hike over a bank of buckled ice, stopping at a long, steepled fracture running parallel to the shore. Lukasi squats for a closer look at a jagged opening. Satisfied, he removes his jacket and chisels at the fracture for half an hour until he breaks through, shaping a hole about twenty inches in diameter. He wastes no time, no strokes. The opportunity to go below opens a couple of hours before low tide and snaps shut when the tide comes in four hours later. This is a window no one wants to miscalculate.

Lukasi drops the ten-foot ladder into the hole and again asks me to wait while he inspects the cavity. His caution, like that of any good hunter, is reasonable, but I am not at all prepared for the feeling of being left suddenly alone while my guide vanishes down a dark hole. I take in, hesitantly, just how alone I am and just how vast, beautiful, and deadly the frozen landscape is. The bay's surface appears quiet and still, but in fact the tide keeps it in constant motion, breaking the silence now and then with ghostly moans and jolting explosions, like gunshot, as the ice adjusts to the ever-changing conditions below.

Before I can think about this too much, Lukasi bobs back up through the hole. "It's good," he says, "lots of mussels and room to forage." He kneels beside the hole and with his knife etches a triangle on the ice's surface to represent the cave's cross section. "There are two important rules to remember when you're under the ice," he says. "Stay where the ceiling is highest; that way if the ice collapses we'll have a chance of getting out." Accidents *do* happen, he told me. Several years ago a young woman was trapped by shifting ice in a cave, or *quuluniq*, and was lucky to get out alive.

"The second rule is not to make loud noises," Lukasi continues. "The vibration can cause the ice to collapse." This second rule intrigued me as much as it scared me, and I decided I would find out more about loud noises under the ice.

Lukasi climbs back down the hole and I position myself to follow, but the hole's too small for my thick layers of clothing. I back out while Lukasi pulls the ladder into the cave. He tells me to use the footholds he carved in the hole. I can't do this either, as there's no room to bend my knees. Lukasi, either indifferent to my conundrum or distracted by the plentiful mussels, disappears into the cave.

It dawns on me that the only way down is to slide. So I do. I let go and drop, arms at my side, into the darkness.

I once parachuted from a plane three thousand feet above Penn-sylvania, and the feeling is almost exactly the same. There's a moment before you jump when your mind must accept that *this is the end*, because in that moment there is *nothing* that tells you that this is safe. It's completely irrational, yet you jump anyway. And, fortunately, above the green meadows of Pennsylvania, the parachute opened.

And, fortunately for me right now, I land six or seven feet below in a pile of ice and rock and seaweed. It wasn't pretty, but I made it down inside the cave.

For several minutes I am completely disoriented by this plunge from a bright, frigid upper world into the dark, warm underworld. My glasses and camera fog instantly. The air is close to sea temperature, forty degrees Fahrenheit—nearly sixty degrees warmer than above the ice. As my eyes adapt to the dimness, I notice the tentlike cavity has a five-foot-high center ridge that parallels the shoreline. The sides taper out like a tent, leaving a forage area of about twenty feet wide and one hundred feet long. The seafloor drops in places, allowing enough head room for me to stand upright. Muted blue light penetrates from above where the ice is thin or fractured. Large drops of meltwater hang from the ceiling, warping like drips from a slow-leaking faucet. The heavy drops slap the bouldered seafloor as they fall, filling the cave with a chorus of snapping fingers. Bits of seaweed and detritus cling randomly to the ceiling—evidence of the last high tide. A crab skitters across a rock and disappears under the seaweed.

As I adjust to the eerie surroundings, my breath shallow and quick, I feel as if I have dropped into an entirely unknown and unexpected realm, as if the tiny hole carved by Lukasi allowed us to slip mysteri-ously, not just under the ice, but beneath the surface of the sea. Here, in a dreamlike state, I feel *inside* the body of the ocean.

～

I first learned about Ungava Bay's tides in 2001 while reading about the Bay of Fundy in eastern Canada. At that time the upper bay, with a documented maximum tidal range of fifty-four feet, six inches, was considered the world's largest tide. There was a notion afoot, however, that the tides of Leaf Basin—around the corner from Lukasi's village—had a tide as large or larger. I was intrigued and wanted to learn more.

I searched the Internet, wrote letters, and made phone calls looking for someone to talk to about Leaf Basin tides or, better still, serve as a guide if I made the trip up there. However, as with many of my travels to outlying destinations, connecting with people in advance proved almost impossible. I eventually decided just to go and see what would happen.

The closest settlement to Leaf Basin is Tasiujaq, an Inuit village of 200 people and 250 sled dogs, so I booked a room in its only hotel (a trailer, actually) and flew aboard progressively smaller planes from Vancouver to Montreal, Montreal to Kuujjuaq (the hub of Nunavik), and finally from Kuujjuaq to Tasiujaq—in a cargo plane choked with cases of Coca-Cola and wine.

I stayed for a week, wearing a hat and mosquito netting each time I ventured from my hotel. On my first walk outside of town, in the midnight dusk of July, I encountered a beluga whale's head. The whale, caught that day, had already been cut into slabs and stored in the community freezer. The head, with skin as pink as a baby's, had been hollowed out and discarded amid the brown-green tundra.

I accompanied hunters and fishermen on treks across the velvety tundra and on boating excursions into Leaf Basin. I was fascinated by the tide's coming and going, which swelled to the doorstep of Tasiujaq and then slipped silently from sight, leaving miles of exposed mudflats, much as it does in the Bay of Fundy. What was it that made these tides so large? I knew about the forty-nine-foot megatides in Bristol Channel, England, and Mont Saint-Michel, France; the forty-five-foot tides in the Penzhinskaya Guba in the Sea of Okhotsk, Russia; and the forty-foot tides in Cook Inlet, Alaska—all record-breaking tides for those countries. I was also aware of the giant tides in Patagonia,

Panama, northwestern Australia, New Zealand, and a handful of other places. But the tides of the Bay of Fundy and Ungava Bay are easily five feet larger than the next largest tides. In my search to find out why, I discovered that the phenomenon which is capable of transforming breath into music and can vibrate glass, bridges, buildings, and airplane wings to the point of collapse is the same force that excites the planet's oceans into a dance we call tide. In all its varied and familiar forms, this phenomenon, called *resonance,* is the tendency of something to vibrate when stimulated. Many factors create exceptionally large or small tides, but none are more significant than the presence or absence of resonance.

On the way home from Tasiujaq, my plane stopped in Kuujjuaq. I was eating lunch at the only café when I overheard a man describing how he hunted under the ice. It was Lukasi. When he finished his conversation I introduced myself. I'd heard of chopping a hole in the ice to fish, to hunt seals, or even to scuba dive, but never to forage for mussels. I later learned that this happens only in Ungava Bay, and only near the villages of Kangiqsujuaq, Saluit, and Quaqnaq.

Lukasi sat with me for half an hour. When we parted, he flying north and I flying south, I knew I had found the guide I was looking for. Best of all, he had invited me to accompany him under the ice that next winter. I was elated. And a little nervous.

I called Lukasi once a month the next winter. He was seldom home, usually out hunting and fishing. When I did catch him, the conditions weren't right—too much or too little ice, not enough tide, too cold, too warm. Finally, a year and a half after we first met, the conditions lined up. Over the phone, with a satellite delay, Lukasi simply said, "Come now."

I had known it would happen that way, so I had a box of rented arctic gear ready. As before, I flew to the Arctic in progressively smaller planes, but the views were markedly different. Where the summer landscape had been pocked with bogs and low-elevation brown-green hillocks, now, in midwinter, it was blanketed in white. North of

Kuujjuaq there are no trees, just low scrub and tundra, all of which disappears under the year's first snow. Off the coast, icebergs miles wide drifted on the black black waters of Hudson Strait. I pressed my face to the frosted window. It all looked so foreign, severe. In a few days I would be following Lukasi into a world beneath what I was seeing.

~

As my eyes adjust to the dim, sultry realm below the ice, I see Lukasi's silhouette bobbing deep in the cave, already picking mussels. I haven't moved after sliding down the hole, which now casts a narrow cone of white light around me. The air is thick. I taste the saltiness. Below, in one of the seaward-reaching furrows, the incoming tide laps against a large seaweed-covered boulder. Lukasi seems to pay no heed, but I'm gripped by an impulse to do some calculations. According to an old mariner's formula, the "Rule of Twelfths," the tide rises at a varying pace. It starts slowly, rising 1/12 of the total range in the first hour, then rises faster in the third and fourth hours (3/12 the total range in each of those hours) before tapering off again.

According to the tide chart I studied this morning, we're in the first hour of the flood. With a thirty-five-foot tide, that means the water in our cave is rising about three feet per hour (1/12), or a foot every twenty minutes. An hour from now, it will be rising at twice that speed. And two hours from now, our little cave will be flooding at nine feet an hour, or eighteen inches every ten minutes. That's faster than a bathtub fills—almost two inches a minute. At that rate, it won't be long before the ice starts to float and shift, transforming our womb-like *quuluniq* into a death trap.

I trust that Lukasi knows the limits, but I can't ignore the nagging fact that the water has already swallowed the boulder it was lapping just minutes ago.

Guided by Lukasi's unflappable confidence, I pull off my gloves and probe under the rockweed where clusters of hard, sharp mussels are plentiful. With a twist, they easily tear from their holdfasts. I fill a

The author and Lukasi picking mussels under the ice.

two-gallon plastic bag in ten minutes without much effort. Lukasi collects twice that in half the time, eventually resting on a rock.

"I was eight years old when my mother and father first took me under the ice," he reminisces, "and at sixteen I was allowed to go alone. My parents taught me, and their parents taught them. We've been doing this for many generations." Lukasi tells me they are able to forage like this only in late winter, when the surface ice is thick and stable. For several days either side of the new and full moon, the tides drop low enough to uncover fresh mussels that are ordinarily submerged. The exceptionally high tides that also occur during these periods tend to shake up the intertidal zone, forming new cavities with room to forage for freshly exposed mussels. Colder winters—something Lukasi hasn't seen for many years—spawn thicker sea ice that can support itself over larger spans, creating more spacious cavities below. As a boy, Lukasi remembers foraging under twenty-foot-tall canopies of ice with his mother and father, getting in and out with long ladders and lighting the way with whale oil lanterns.

Born in 1949 in a snow house at the base of Ukaliq (Arctic Hare) Mountain, Lukasi is part of the last generation to remember traditional ways of life that are nearly lost. Before packaged foods were available in the 1960s, starvation was an ever-present possibility, and collecting mussels under the ice was necessary for survival. Some families, including Lukasi's, migrated to Kangiqsujuaq because of this reliable food source.

The 1950s and '60s brought a quantum leap in military, scientific, and industrial interests in the far north, and with this came an invasion of people, goods, ideology, and diseases. Lukasi's generation knew the last village shaman, watched the construction in the early 1960s of the first hospital and airport, and, in 1975, witnessed the signing of the James Bay Agreement, an Aboriginal land-claim settlement that brought subsidized plywood houses, sewer systems, electricity, water, and schooling to the fourteen villages of Nunavik. Lukasi saw himself in a mirror for the first time at age twelve.

To the horror and disbelief of the Inuit, a controversial "Dog Law" initiated by the Royal Canadian Mounted Police resulted in the destruction of 10,000–20,000 sled dogs between 1950 and 1970, a campaign ostensibly launched to control strays. The campaign hastened the transition to snowmobiles for winter transportation. The first general store, the Nunavik Co-op, was built in 1964 by the Makivik Corporation, an Inuit-run entity that looks after Native interests in northern Quebec. The co-op has a store in each of the Nunavik communities, where members can buy guns and ammunition, snowmobile parts, clothing, tires, diapers, bacon, soda pop, and Cheez Whiz. Though Lukasi managed his village's co-op for thirty-five years and buys half his family's food there, he still refuses to buy or eat an animal packaged without its head. "I want to know what I am eating and where it came from," he insists.

Lukasi pulls another plastic bag from his pocket and resumes mussel-gathering. The sea level has risen a couple of feet, and I wonder how much longer he will allow us to stay. In no apparent hurry, he crushes a mussel against a rock and uses a broken shell to clean the beard from another mussel. With a few deft strokes, he splays the bivalve on his open palm, scoops out the soft yellow body, and pops it in his mouth. He prepares another for me. "Sometimes we use a *quuluniq* to hide from bad weather," he says as we chew the sweet, springy meat. "And my wife and I have enjoyed eating lunch and cooking tea while picking. I am teaching my children and other young people in the community to hunt under the ice, but many of them are afraid or not interested. They'll play hockey and basketball or travel to the big cities of Quebec or Montreal, but when it comes to learning subsistence skills, they're losing interest. My father got three-quarters of his food from the land, I get half of mine, and my children get a quarter of theirs."

Lukasi's transition from animistic spiritual values to Christianity has not been seamless. At the fringes of what appears to be a fully adopted Christian life, there are signs of the old ways. Although he attends church and says prayers before feasting on fresh mussels, he refers admiringly to the shaman's power. A good shaman, he tells me, had the ability to retrieve a lost knife from the bottom of the sea. The younger generation is less ambivalent, yet they struggle to reconcile the past with the present. When I asked the museum curator, a young woman in her early twenties, if she viewed the world as her ancestors did, where everything—every rock, every tree—is imbued with spirit, she answered curtly, "My spiritual beliefs are a private matter between me and Christ."

Lukasi puts a couple of more mussels in his bag and stands. "We should go now," he announces, and I leave no gap between us as we scuttle toward our escape hole. He climbs the ladder first and, once up top, widens the hole a little so I can squeeze through.

Above, the cloud cover has lifted, and shafts of white evening light cast long, haunting shadows. On the other side of the bay, a several-hundred-foot granite face ascends vertically from the water.

The tide rises and falls against this face, depositing a curtain of polished blue ice. At low tide, this thirty-five-foot-tall curtain stretches for miles along Wakeham Bay's west shore.

As the sun sets, the landscape grows ever more formidable. "This time of year," explains Lukasi, "the temperature drops so fast after sundown that a person will freeze to death in a matter of hours." Several days ago, two men left their village in the morning on a hunting trip, expecting to return before dark. Apparently their snowmobile flipped late in the day, injuring one of them so badly he couldn't walk. The other man left on foot to look for help and stumbled into Kangiqsujuaq after sunset. Within minutes the village had rallied a search team. "The only way to survive is by building a snow house, which the injured man was unable to do," Lukasi says. The search was made more urgent by the arrival of a weather system, bringing wind and new snow. Such conditions could make it impossible to follow tracks, which are often the only link between the frozen wilderness and the safety of the village. Luckily, the man was found, weak and disoriented, at eleven o'clock that night.

We load the sled and I climb back aboard the snowmobile behind Lukasi, clasping fistfuls of his jacket and thinking about those nearly lost hunters as we glide back toward the village. Evidence of the rising tide is everywhere: stress fractures have opened up like veins, and the once-crumpled ice of the intertidal zone is lifted and floating. As we near the bay's head, the boxy houses of Kangiqsujuaq, painted red and blue and yellow and suspended on spindly posts, are a comforting sight.

〜

Lukasi and his community wouldn't be able to hunt under the ice if it weren't for this region's enormous tides. For all the changes imposed upon northern indigenous cultures since European contact, what has not changed at all is the coming and going of the tides. Since the Inuit first arrived in northeastern Canada more than two thousand years ago, these tides have played a vital role in shaping their way of life. During spring, summer, and fall, when the sea ice breaks up, the Nunavik communities take advantage of low tides to gather mussels, clams, crabs, and snails. In most villages, which are settled around a protected bay, the tide recedes so far that it disappears entirely, uncovering miles of land. In just a few hours, a view that was once filled to the horizon with shimmering blue water is transformed into a moonscape. Small boats are left high and dry, listing on their chines. There are no marinas or boat launches except where the water is deep enough, usually at the bay's mouth, miles from the village. At low tide, monofilament gill nets are anchored to shore and stretched far into the intertidal zone. As the tide rises, the nets are suspended by small floats—usually plastic pop or detergent bottles—snaring arctic char. At the next low tide, the nets settle back on the gravelly bottom where family members take turns harvesting the char. Growing to thirty pounds, these bright, muscular fish are a prominent part of the summer diet, along with caribou, seal, musk ox, and intertidal shellfish.

With no roads connecting the villages, goods are shipped in by small plane or, during the summer, by freighters originating in Quebec or Montreal. Loaded with vehicles, fuel, staples, and prefabricated house parts—*everything* needed for the long winter—freighters come only once or twice a summer, when the tide is right. They must schedule their arrivals and departures at each of the fourteen Ungava Bay villages to coincide precisely with the highest tides; otherwise they can't offload their supplies. The transfer of fuel from ship to barge and from barge to shore-based tank is a feat involving an exacting choreography of tide, time, and navigation. The barge, laden with a full winter's supply of fuel, must nose in to shore, hook up, unload, and get out while there is still water enough to float it. One false move—a delay, a botched transfer—could cause the barge to go aground, potentially tearing open its hull and spilling fuel. Boulders in the intertidal zone shift around with the tides, adding further hazard.

A summer freighter arrived at Tasiujaq when I was visiting in mid-July. It was a festive occasion. The whole village knew that the full moon meant large tides and the arrival of long-awaited goods. When the day came, every soul in the community gathered waterside at high tide to help offload the barge. Dogs howled, forklifts whirled, children watched expectantly. Someone was waiting for a new snowmobile; someone else for new winter clothes; others for hunting gear, boat parts, or furniture. The flurry came and went with the tide; several hours later, the water and the ship had disappeared and the village was once again quiet.

～

After my sojourn under the ice, I begin puzzling over Lukasi's second cardinal rule for the ice cave: no loud talking. His concern is that a loud voice might resonate in the body of the cave—much like a plucked string resonates in the body of a guitar—and cause the ice to shift or, worse, collapse. What I hadn't expected to learn is the resonating effect of loud voices is the same phenomenon that creates Ungava Bay's large tides.

In the seventeenth century, Kepler and Galileo had hinted at the role of resonance in shaping the tides. In his book *The Harmony of the World*, Kepler recognized a harmonic relationship between the planets and intimated that these harmonies worked together with powerful effect. Galileo, as we learned earlier, observed water sloshing in jugs aboard a Venetian boat and conjectured that each ocean had a natural period of oscillation that would cause it, like the jugs of water, to slosh back and forth with a unique rhythm. A quarter-century later, Laplace, Airy, and the *Beagle*'s Captain Fitzroy all wondered too whether the tide, rather than a progressive wave, was each individual ocean's response to the tide-generating forces in the heavens.

By Laplace's time—the late eighteenth century—astronomy had become an exact science. It was known, for example, that many astronomical cycles affected the tides. The obvious ones were the moon's monthly change from new to full and back again, the near-monthly cycle of perigee and apogee, the moon's changing declination north and south of the equator, and the seasonal change in the earth's declination relative to the sun. Less obvious was the varying distance of the earth from the sun, closest in January (perihelion) and farthest in July (aphelion). Longer cycles were also known, such as the 8.8-year cycle in which the closeness of the moon during perigee varies, and the 18.6-year cycle of the moon's declination (its maximum monthly declination varies from 18.3 degrees to 28.6 degrees).

To make sense of all these cycles—many of them nested within each other—and their effect on the tides, Laplace separated them and treated each as an isolated component. He then imagined a stationary earth with these tide components circling as satellites. Hence, the lunar day was represented by an imaginary satellite that circled the earth in twenty-four hours and fifty minutes. To account for the tide bulge on the opposite side of the earth, Laplace added another moon satellite circling in twelve hours and twenty-five minutes. These two satellites represented the moon's daily tidal influence. Laplace added imaginary satellites, what he called *astres fictifs* ("fictional celestial bodies"), for all the known tide-generating cycles, and put them roughly into daily and twice-daily groupings. This approach became the framework for modern harmonic tide theory, which was picked up and developed further by the British mathematician and physicist William Thomson (Lord Kelvin) in the 1860s.

Thomson is best known for his work in thermodynamics (low temperature scales and degrees Kelvin), but he also made contributions in electricity, magnetism, the design of undersea cables, and improving the marine compass. On behalf of the British Association for the Advancement of Science, Thomson established and chaired a committee in 1867 "for the purpose of promoting the extension, improvement and harmonic analysis of tidal observations." Outlining his approach, he wrote: "The height of the water at any place may be expressed as the sum of a certain number of simple harmonic functions of the time, of which the periods are known, being the periods of certain components of the sun's and moon's motions. Any such harmonic term will be called a tidal constituent."

Once all the tide-generating forces were isolated and individually represented by fictional satellites, it then became a matter of analyzing the ocean's response. How did the Atlantic, for example, respond to the lunar cycle of twenty-four hours and fifty minutes? How did it respond to the lunar cycle of twelve hours and twenty-five minutes, or the sun's twelve- or twenty-four-hour cycle? This, as Laplace and others had speculated, was much more complex than it at first appeared. As Galileo suspected, each ocean does indeed have its own natural period of oscillation that predisposes it to respond energetically to some celestial pulses and not to others. To understand this, we have to understand resonance—the reason Lukasi cautioned me not to make loud noises under the ice.

We experience acoustic resonance every time we sing in the shower and hit that sweet spot where our voice is suddenly and richly amplified by the shower stall's vibration. We tune into our favorite radio station by rotating a knob that allows our receiver to resonate with just one of hundreds of signals. The stronger the resonance, the clearer the radio reception. Musical instruments of all kinds—wind, string, percussion—rely on a resonating body to create and amplify tones. Concert halls, too, are meticulously designed and constructed to resonate with the music played in them. Chartres Cathedral in France is so vibrant that a composer once exclaimed, "This place is more alive than I am!"

Another familiar experience of resonance is a child on a swing. The swing has a natural rhythm that can be accentuated by a push at the right time or diminished by a push at the wrong time. The "right" push pumps more energy into the system and causes resonance, sending the swing higher. The "wrong" push is out of sync with the swing's natural rhythm and causes it to slow or stop. Resonance also occurs when our car vibrates annoyingly—and sometimes uncontrollably—as we drive down a washboard road at a speed perfectly synchronized with the bumps. We learn to avoid this by speeding up or slowing down.

On *Crusader*, the pilothouse windows start rattling when the engine reaches 2100 rpm; the rattling stops if I speed up or slow down.

Resonance is everywhere, even in unexpected places. Kim Bostwick and a science team from Cornell University learned that a male club-winged manakin, a rare South American bird, uses resonance to attract a mate. The sparrow-size manakin has nine inner-wing feathers, the innermost of which have enlarged hollow shafts. When rubbed together at the right frequency, these innermost feathers excite all nine feathers to resonate in unison, wooing females with a sustained violin-like note close to F-sharp.

Oceans, like swings and feathers, have inherent rhythms that can be excited by a resonant push from the celestial world. To apply the example of the shower, the stall is the ocean and our singing is the gravitational influence of the sun and moon. The "sweet spot" occurs when the sun or moon hits a "note" that stimulates a sympathetic vibration in the ocean, which in turn amplifies the tide.

This excitation introduces a new dimension to our thinking about tides. Where we once thought the sun and moon pushed and pulled the oceans around the planet, we now understand that it's a partnership. The sun and moon call out, as it were, and the oceans call back. The oceans aren't passive listeners but partners in an energetic conversation—resonance—that ultimately accentuates or diminishes the tide.

The sun and moon's call is the gravitational force generated by any one of hundreds of orbiting cycles. Depending on their resonating characteristics, the oceans call back, sometimes with enthusiasm and sometimes with apparent lack of interest. Like a mother who singles out her child's voice among the many voices in the playground, some ocean basins are so finely tuned to one astronomical signal that they don't "hear" the others.

⚏

Six months after my Arctic visit, I'm outside a hotel in Nova Scotia, on the shores of the Bay of Fundy, waiting for Charlie O'Reilly. A retired chief tide officer for the Canadian Hydrographic Service, O'Reilly has just returned from a ten-thousand-mile motorcycle trip to southern Baja, Mexico. He's agreed to give me a tour of the Bedford Institute of Oceanography and will be taking me there on his Honda Shadow. It's summer. The day is warm and dry. When he arrives, O'Reilly is decked in black leather. He removes his helmet and greets me in the same affable manner I've sensed during our many phone conversations. He is stocky, with short gray hair and deep-set, lively eyes.

As we rumble down the highway, my hands gripping O'Reilly's waist, I think of how just six months ago I was doing something similar with Lukasi, only on a snowmobile. There are lots of differences, of course, but in both cases I had my arms wrapped around a man I hardly knew, a man who generously agreed to be my guide, a man to whom I entrusted my life. One is a hunter who knows the tides through experience; the other is a scientist who knows the tides through mathematics and modeling.

When we arrive at the Bedford Institute in downtown Halifax, O'Reilly takes me to the security office, where my daypack is searched and I'm given an identification card. Inside, we wander through wide, vinyl-tiled halls. Glass cases display maps, photos, and three-dimensional models of the bay's oceanographic history.

When I ask about resonance, O'Reilly stops and describes a circle with his hands. "Imagine this is a table," he says, his voice resounding in the hall. "Now put pans on the table—many of them, each a different size and shape, each filled with water to different levels.

"If no one jostles the table, the water in the pans will sit still. But if someone starts kicking the table's leg," he exclaims, as he mimics a kicking motion, "the water responds. Some of it is excited into a lot of motion, and some of it hardly moves at all. That's how resonance works in the ocean."

To better understand O'Reilly's example, it helps to remember that everything in the universe has a natural tendency to vibrate: flowers, wind, steel, planets, mountains, the inside of an ear. "*Everything* at its most microscopic level," writes physicist Brian Greene, "consists of combinations of vibrating strands." If we could trace the pattern of these vibrations, they would look like waves, some short and steep, some long and gradually sloped, some traveling slowly, and some fast. The frequency of these waves, or the number of peaks that pass a given point per second, varies. Radio and television signals vibrate at 20,000–150,000 peaks, or *cycles*, per second. The universal reference note of middle C vibrates at 253 cycles per second; a hummingbird's wing vibrates at about 20 cycles per second; earthquakes vibrate at 1–5 cycles per second. All these are high-frequency vibrations.

The ocean, sun, and moon vibrate at much lower frequencies, measured not in cycles per second but in *periods*, or the time between the passage of successive wave peaks (or pulses). An average tsunami, for instance, has a period of about one hour; the solar day—the sun's equivalent of a wingbeat—has a period of twenty-four hours.

Anything that vibrates has what's known as a *natural period of vibration*, the frequency at which it vibrates if simply kicked and left alone. A bell is a classic example. When it's struck just once, it vibrates at a natural frequency defined by its shape, size, and weight as well as the material from which it's made. Change any of these defining characteristics, and the natural period of vibration will also change, giving the bell a different pitch.

Oceans too have a natural period of vibration, what Galileo and modern oceanographers refer to as the *natural period of oscillation*. When excited, oceans oscillate back and forth at a characteristic frequency, like a wave in a bathtub. The Pacific Ocean has a natural period of oscillation of about twenty-four hours, the Atlantic about thirteen hours, the Antarctic about thirty hours, and the Baltic about twenty-seven hours.

If it were possible to give the Atlantic a big kick, it would slosh back and forth in its basin like waves in a bathtub, peaking first off the coast of France in the north and then, thirteen hours later, off the coast of Patagonia in the south.

The force that causes something to vibrate might be a single episode, like an imaginary kick to the Atlantic or strike of a bell. In such cases, the responding body sloshes or vibrates at its natural frequency, and the vibration slowly fades due to gravity and friction. If, however, the force is applied at periodic intervals, and if those intervals match the natural frequency, then the result is infinitely more complex—and magical. A single push of the swing from Mom and Dad is one thing; a synchronized push over an extended period of time is entirely different.

Imagine what would happen if the Atlantic, with a natural frequency of thirteen hours, received not just one kick but many kicks over a sustained period of time and that each was delivered every thirteen hours, just as the sloshing wave peaked in the north or south. This action is precisely what happens when we push our child on the swing at the "right" moment over and over again: the swing resonates and the child giggles with joy as she soars. In the Atlantic, it's the moon's push that happens most strongly at the right moment over and over. In the Pacific, it's the sun's. If the pushing, or forcing, occurred at a significantly different time interval, there would be no resonance, no synchronicity, no giggling.

The table-kicking that O'Reilly describes is delivered by any one of four hundred different astronomical cycles. These are what Laplace called fictional celestial bodies and what Lord Kelvin called *harmonic constituents*, the name used by scientists today. Some of these constituents issue a walloping kick; others are hardly perceptible nudges. Some have very, very low frequencies, such as the 25,800-year wobble of the earth's axis called the precession of the equinoxes. Some have higher—and more influential—frequencies, such as the daily and twice-daily cycles of the sun and moon.

Of the four hundred tide-generating constituents, only twelve are responsible for 90 percent of the tide, 90 percent of the time, which usually renders a prediction accurate to within several inches—good enough for coastal navigation.

Most countries use more than twelve, however. Germany uses

twenty, the United States thirty-seven, England and Canada sixty. In waterways needing greater precision, more constituents are considered. At Alaska's Cook Inlet, with large tides and heavy oil-tanker traffic, the National Oceanic and Atmospheric Administration (NOAA), the entity responsible for publishing the U.S. tide tables, considers 120 constituents. These constituents are fed into supercomputers— machines, that is—which spit out yearly predictions in minutes.

Most of the twelve primary constituents are the daily and twice-daily "kicks" from the sun and moon. The cycle that has the biggest periodic kick in most ocean basins is the moon's semidiurnal cycle of twelve hours and twenty-five minutes, known as M2 (moon, twice daily). This cycle is exactly half a lunar day, representing the moon's appearance overhead on one side of the earth and then, twelve hours and twenty-five minutes later, on the opposite side.

The sun's semidiurnal cycle of twelve hours, known as S2 (sun, twice daily), is exactly half a solar day, and it also exerts a strong influence on ocean basins that vibrate close to that frequency—the Pacific, Antarctic, and Indian Oceans, among others. The Pacific Ocean is unique in that its natural frequency of twenty-four hours is matched perfectly with the sun's twenty-four-hour cycle. This is why the tides in the Pacific have a strong diurnal (once daily) component.

≈

Two other important facts must be added to O'Reilly's analogy. The first is that there are countless ocean basins, or pans, on his imaginary table. When we think of the world's oceans, we usually think of the five largest: the Pacific, Atlantic, Indian, Arctic, and Southern. We might readily add a few smaller seas: the Red and Black, Caribbean, Mediterranean, and Baltic. The *World Atlas* identifies hundreds more: the Bering, Beaufort, and Greenland Seas; the Caspian, Adriatic, Arabian, Barents, Kara, Norwegian, Chukchi, Yellow, Balearic, and Ionian Seas; the Bay of Bengal, the Sea of Japan, the Gulf of Aden, and so forth.

The boundaries of these basins shown in the atlas are often political and arbitrary, but to physical oceanographers, water boundaries have nothing to do with politics and are *anything* but arbitrary. Defining basin boundaries is, in fact, critical to understanding how the tides work—and particularly how resonance works.

A body of water, whether a small lake or the Atlantic Ocean, may appear on the surface as one basin, but below the surface there may be features that divide it into multiple basins. These divisions are defined by differences in water density, temperature, salinity, and, most commonly in the ocean, seafloor topography.

The deep regions of large oceans usually behave as one basin, or pan, on O'Reilly's table. On the shallow continental shelves, there are many basins, even though their distinction is not detectable at the surface. The Bay of Fundy basin, for example, stretches south to the Gulf of Maine, north to Newfoundland, and out to the continental shelf's edge.

This Fundy–Gulf of Maine system responds as one pan on O'Reilly's table, and it has a natural period of oscillation of 13.3 hours—close to the 12.47-hour pulse of the dominant moon cycle, M2.

Ungava Bay is part of a pan that includes Hudson Strait and, like the Fundy–Gulf of Maine system, stretches out to the Atlantic shelf. This system has a natural vibration of 12.7 hours—also closely synchronized with M2.

Narrow channels, too, can create boundaries between basins. The Strait of Gibraltar, for example, effectively isolates the North Atlantic from the Mediterranean. The tides on the Atlantic side of the strait are twice as high as those just fifty miles inside the Mediterranean. The Skagerrak, a narrow, wending channel between the North Sea and the Baltic, accomplishes a similar separation. With a resonant period of 30 hours, a near multiple of the primary lunar and solar cycles, the North Sea resonates strongly, whereas the Baltic, with a natural period of 27.3 hours, hardly resonates at all. In fact, the seasonal outflow from coastal rivers contributes more to the two-foot rise and fall at Malmo, Sweden, than do the sun and moon.

Ocean basins can resonate with "kicks" from sources other than the moon and sun. The Adriatic resonates with African winds, called siroccos, that blow during the spring and fall. In Venice, at the Adriatic's north end, these hot, dusty winds can excite a three-foot rise in sea level. Small harbors, too, are known to resonate with subtle and unpredictable vibrations felt at their entrance, such as the pulse of breaking waves or subsurface currents. Known as *harbor effect*, these responses can't be accounted for in tide charts but are often substantial: the water level in Los Angeles Harbor can unpredictably rise or fall ten feet due to this effect.

In shallow water, *overtides* come into play. Analogous to musical overtones, overtides include higher frequencies that are multiples of the primary daily and semidaily frequencies. In other words, these "kicks" can be delivered as often as every six or eight hours. The Patagonia shelf off the southeast tip of Argentina is well known for its energetic response to overtides.

The other piece of information to add to O'Reilly's analogy is that all the table pans are touching, and hence the resonance of one affects the resonance of all. The pan on the continental shelf that we just separated from the neighboring deep basin is not actually separate, nor is the Atlantic from the Mediterranean, or the North Sea from the Baltic. Although these ocean basins are treated as separate entities for the convenience of study, in reality they are inextricably connected. The table-leg kick delivered by the 2002 earthquake in Indonesia showed up as a measurable rise in sea level in the Bay of Fundy, a vibration that traveled through the Indian Ocean, around the south end of Africa, and up the full length of the Atlantic—covering more than thirty thousand miles in fifty hours—to lick the red sand cliffs of Burntcoat Head, Nova Scotia.

That same earthquake sent out a ground vibration that resonated with thousands of swimming pools and lakes within a five-thousand-mile radius, causing some to overflow. The Ungava Bay–Hudson Strait system is excited into a lot of motion by its marriage with the Atlantic but, as in any good marriage, the Atlantic is also excited by Ungava Bay, which exaggerates the tides all the way to the coasts of France and England. Oceanographers call this *back effect*.

It's impossible to overestimate the importance of the linkage between and among basins—also known as *coupling* or *co-oscillation*. In many cases, this linkage is the primary driver of the tide. Lake Michigan, with its natural oscillation period of thirteen hours, is resonant with the moon's M2 forcing, yet it has a tide of only three inches. This is because Lake Michigan is an isolated body of water—a pan by itself. The Bay of Fundy and Ungava Bay, like Lake Michigan, have resonant periods closely aligned with M2, but the exceptional tides in these basins are more a result of coupling with the Atlantic than an independent response to a "kick" from the moon. Take away the Atlantic, and the extraordinary tides of Fundy and Ungava become ordinary. It is fair to say, then, that the moon moves the Atlantic and the Atlantic moves Fundy and Ungava. The deep Atlantic, like most large ocean basins, has tides of only one to two feet, but this huge body of water packs an immense amount of energy. The pulse it delivers to the coastal pans may be small in height, but it's massive in strength.

⁓

Resonance is often magical, but it can also be ruinous. It's not a myth that marching soldiers break step when about to pass over a bridge, for fear their rhythm might resonate with the structure and shake it to pieces. In the Angers bridge catastrophe of 1850, 478 French soldiers marched in lockstep across the Basse-Chaîne Bridge in Angers, France. The bridge collapsed, killing 226 men. This was not the first or the last bridge to collapse from resonance. A sign on the Albert Bridge in London reads, "All troops must break step when marching over this bridge." In the late fall of 1940, the Tacoma Narrows Bridge in Washington State shook apart in sustained forty-knot winds. The bridge was nicknamed "Galloping Gertie" because, even during construction, it pulsed up and down when resonating with only modest breezes.

Buildings tend to have natural oscillation periods of one to six seconds, which is unfortunately matched to the ground vibration of most earthquakes. The reason Hawaii's Hilo Bay suffers so much damage from tsunamis is that it resonates at thirty minutes—a perfect match to (or a close multiple of) the vibration period of most tsunamis. The resulting resonance amplifies the already large waves in the same way that a person sloshing in a bathtub, if he does it at the right rhythm, can build up a wave that jumps the tub. The 1960 earthquake that rattled Chile (at 9.5 magnitude, the largest quake of the twentieth century) gave birth to thirty-five-foot-high monsters in Hilo Bay, killing sixty people and destroying hundreds of homes and businesses. Elsewhere in the Hawaiian Islands, these same waves were three to seventeen feet and caused far less destruction.

The resonating personalities of ocean basins can change over time, especially those near the coast that are subject to a variety of geological processes and/or human manipulation. O'Reilly tells me that Fundy's tides haven't always been large. "Fifteen thousand years ago," he says, "we were a mile under the Laurentide Ice Sheet, and sea levels were much lower than they are today." As glaciers retreated, the land rebounded and sea levels rose. Models indicate that seven thousand years ago, the Bay of Fundy had a period of oscillation of roughly eighteen hours and a tide range of ten to twenty feet, and by four thousand years ago it had an oscillation period of fourteen hours and a tide range of about forty feet.

Around 2,500 years ago, sea level rose enough to flood Georges Bank, a large, shallow part of the Fundy–Gulf of Maine system. When that happened, the bay's oscillation period shifted closer to resonance. Today the resonant period is about 13.3 hours—not far from perfect alignment with the Atlantic and the M2 cycle.

Resonant periods can also change due to coastal development projects. In the 1970s oceanographer Chris Garrett studied the effect of a tide-generating dam across the Bay of Fundy, which had been proposed by the Canadian government to offset that decade's energy crisis. Garrett determined that the dam would change the resonant quality of the bay so much that Fundy's tide would get smaller and Boston's would get larger. Canadians teased that it was a perfect supply-and-demand fit: energy from Fundy's tide would supply the electricity needed to pump out flooded basements in Boston.

〜

Until the 1990s, the Bay of Fundy held the title of the world's largest tides, based on a measurement at Burntcoat Head in Minas Basin of 54.6 feet between an adjacent high and low water. This exchange, recorded in 1952, exceeds by five feet the planet's next largest tides: Bristol Channel in England and Mont Saint-Michel in France. Fundy's claim was acknowledged in the *Guinness Book of World Records* and in the *Sailing Directions for the Gulf of Maine and Bay of Fundy*. The three hundred thousand people who live in Nova Scotia and New Brunswick never considered anything different. "When I'm abroad," wrote Joseph Howe, a politician who lived 170 years ago, "I brag of everything that Nova Scotia is, has, or can produce, and when they beat me at everything else, I say, 'How high do your tides rise?'"

But the 9,200 people who live in the fourteen villages that hug the shores of Ungava Bay, 1,200 miles north, *do* think differently, and they raised the question to the Canadian Hydrographic Service (CHS), the official entity responsible for tracking such matters. "There was a contradiction in the *Sailing Directions*," claims Allen Gordon, director of the Nunavik Tourism Association, based in Kuujjuaq. "For Fundy, the *Sailing Directions* listed Minas Basin as the world's largest tide, and the same publication for the Arctic region listed Leaf Basin in Ungava Bay as the world's largest tide."

In 1997 Gordon wrote a letter to CHS citing the contradiction. Charlie O'Reilly, the chief tidal officer at the time, responded that because of budgetary constraints and because the issue was not a matter of navigation safety the service was unable to fund a study to resolve the contradiction.

O'Reilly referred Gordon to a private consultant, and the next year, in 1998, three tide-measuring gauges were placed in Leaf Basin near Tasiujaq. The result was a recorded tide of 50.6 feet.

After several errors were discovered in the testing process, the gauges were set again. "The more we looked at it," said Gordon over the phone, "the more complex it became. If the tide difference between these two places were five feet, it would be easy to determine the winner. But because they're within *inches* of each other, the winner can't be decided without a high level of precision in methodology and technology, which we don't have."

In 2001 a third round of gauges were set, revealing a 55.84-foot range. "The consultant's report," said Gordon, "described the difference between the tides of Ungava Bay and Fundy as 'so small that the two may be considered equivalent.'" In that report he also noted a seven-inch margin of error in the gauges, which is enough to skew the outcome one way or the other when measured on any given day.

The people of Nova Scotia and New Brunswick have not taken the Ungava challenge lightly. When news of it appeared in the press, a flurry of letters to the editor expressed disbelief and, in some cases, outrage.

Journalist Will Ferguson described an encounter at a New Brunswick party: "As I stood crowded in beside the fridge, I mentioned, just in passing, that the tides along the Bay of Fundy might not be the highest in the world after all. . . . Well, you would have thought I had peed on their rug. Not only was I shouted down and threatened with physical violence and, even worse, banishment from the kitchen, but soon I began to fear for my life as well. I had angered the gods of Fundy."

In the fall of 2002 the Canadian Hydrographic Service issued a letter, signed by O'Reilly and a colleague, acknowledging the difficulties inherent in measuring tides as large as Fundy's and Ungava's. The letter concludes: "It is therefore appropriate to say that these two sites exhibit comparable extreme tidal ranges to within the current ability to determine these measurements."

In 2004 the CHS changed the *Sailing Directions* for the Arctic region to read, "The tidal ranges in [Leaf Basin] are among the highest in the world." And in 2008 the *Guinness Book of Records* listed both the Bay of Fundy and Leaf Basin as the world's greatest tides.

When O'Reilly dropped me off at my hotel at the end of the day, he admitted to being exasperated by the rivalry. "This has been going on for ten years, and every time something about tides pops up in the news, we're called on to comment. The controversy quiets down and you think it's settled, but then it springs up again. Who cares whether one bay has a tide seven inches larger than the other? It's like having the owners of the two tallest houses in the neighborhood fight over which one's tallest. Really, I get sick of it. Both tides are almost five feet larger than the next largest tides in the world, so what fucking difference does it make?"

〜

During my week in Tasiujaq a couple of years earlier, I joined two Inuit seal hunters for a day trip by skiff into Leaf Basin. Although Leaf Basin is eight miles across and ten miles long, it's protected from ocean swell and usually calm enough for small boats to navigate safely. On the summer morning we launched our eighteen-foot skiff, there was just enough breeze to lightly dimple the water and keep the bugs away. We motored out of Tasiujaq at slack high water and would see no other boats or people until our return with the next flood, a little over twelve hours later. We crossed the bay and skiffed several miles into a fjord. As it narrowed, I studied the sun-washed granite where orange and white lichen, dwarf alder, and small bundles of knotweed filled every crevice. My companions fired their rifle at several ringed seals but missed. Nevertheless, they seemed content to drift, eating crackers and casting leisurely for char as we whiled away the morning.

"In an hour, when the tide turns, you won't recognize this place," the skipper tells me. "It will be transformed into a lethal storm of white water. We'll have to leave soon or we'll be caught." In the exquisite calm, it was hard to believe.

When it was time to leave the fjord, I asked to be dropped off where I could hike up a granite knoll and watch the tide turn. From there, just two hundred feet above sea level, I stood among delicate wildflowers and gazed over miles of iridescent land and sea. Earlier in the day one of the hunters had spoken reverently of *innua*, the spirit of place that inhabits animals, rocks, plants, mountains. "Everything is alive," he said, adding that he saw no difference between himself and the land, the moon, or the tide. "These are powerful forces, and my relationship to them binds me to my community and my ancestors."

As predicted, in less than an hour our peaceful fjord was unrecognizable. The dimpled surface exploded into torrents of white water, just as it does at Skookumchuck, stumbling and pushing its way out to sea. Our skiff, had it been caught, would have foundered.

I had watched the same transformation years ago from tall bluffs at Cape Split in Nova Scotia. From that grassy lookout, where gulls glide on updrafts, I gazed across miles of Fundy's ruddy waters and, as the tide turned, watched the smooth surface erupt. Oceanographers claim that the volume of water that flows in and out of the Bay of Fundy during a tide cycle is equal to the combined outflow of all the earth's rivers during the same period. I imagine the same is true of Ungava Bay. In the face of such magnificence, differences measured by a stick or gauge seem irrelevant.

The resonant calls of the sun and moon have frequencies too low for me to sense, but on that splendid summer day in Leaf Basin I could taste the salt kicked up by the awakened tide and feel its thunder resonating in my chest.

8 Turning the Tide

Grinding Wheat, Powering Homes

If 0.1 percent of the renewable energy available within the oceans could be converted into electricity it would satisfy the present world demand for energy more than five times over.

—*UK Marine Foresight Panel*

When Ferdinand Magellan set sail from Spain in the summer of 1519, he was looking for "the Dragon's Tail," a passage rumored to cut across the southern tip of South America, connecting the Atlantic with the Pacific. Fourteen months later, he was elated to find the passage, but his ships were ill prepared for the violent weather and wild tidal currents of the three-hundred-mile labyrinth that now bears his name. The worst of it was concentrated in an hourglass channel, known today as First Narrows, just inside the Atlantic entrance. With twenty-foot tides and ten-knot currents pouring from the straits, his ships, powered only by wind, were overwhelmed. Although his accomplished Portuguese and Spanish crews were familiar with tides, these were six times larger—and the currents exponentially fiercer—than anything they'd encountered at home. On a typical day, even with a fresh breeze, his ships would make no more than seven knots. Against a ten-knot current, they'd be sailing backward.

In the sixteenth century—the early years of the Age of Discovery— the European map of South America ended just south of Buenos Aires. Beyond, there was nothing—in fact, worse than nothing: the world was still considered flat by many, so ships sailing south on the ocean-sea,

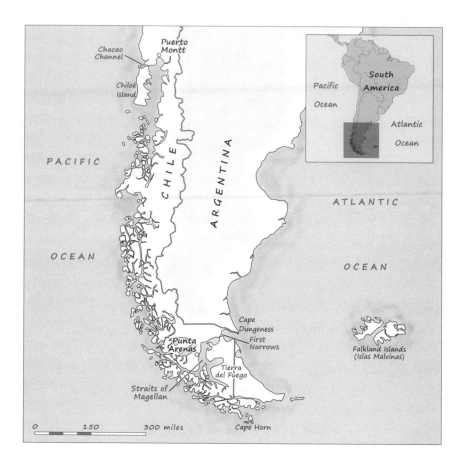

as the Atlantic was called, were embarking not just on uncharted, monster-infested waters but quite possibly on a course that would lead them off the world's edge. No European ship had ever returned from a journey beyond Buenos Aires. Yet Magellan's ambition was to sail through the fabled Dragon's Tail to the Spice Islands in the western Pacific, returning to Spain laden with cloves, nutmeg, cinnamon, and other riches. He was ruthless in this endeavor, often beating his crew or withholding food and water as a means of punishment.

By the time Magellan sighted the narrow opening between present-day Patagonia and Tierra del Fuego, he and his crew had survived months of storms, an attempted mutiny, and the loss of one of his five ships. Antonio Pigafetta, the fleet's chronicler and one of the few to survive the expedition, wrote: "After going and setting course . . . toward the said Antarctic Pole . . . we found by miracle a strait which we called the Cape of the Eleven Thousand Virgins."

Magellan's crew struggled into the Dragon's Tail but were spit out by the current. They had to wait out the ebb tide at anchor and sail with the flood, which pushed them into the straits. This sounds simple enough, but the passage was entirely uncharted, so every move was a risk. Secure anchorages were difficult to find, as the ebb's relentless pull led to one dragging anchor after another.

Inside the straits, weary crew members were fooled by channels dead-ending in the mountains and befuddled by high and low tides that matched no pattern they had seen before or could understand. And for good reason: they were at the very point where two large oceans meet—two oceans whose tidal periods don't match. The Pacific tide usually crests a few hours ahead of the Atlantic, which creates a confusing seesaw effect in the straits. This was nothing like the relatively predictable tidal patterns on the coast of Spain. Exhausted and homesick, Magellan's crew members could only guess what they were up against. For all they knew, the water's turbulence was a sign that they were indeed about to be sucked over the edge of the world.

It took Magellan six weeks to find the "small gulf on the other side"—the planet's largest ocean—which he named the Pacific. As he sailed toward the Spice Islands in search of the exotic cargo that he hoped would bring him fabulous wealth, he could not have guessed that the tidal power that vexed his voyage through the Dragon's Tail would one day not only be understood and charted but harnessed as a source of heat and light.

≈

East of the Andes, where I am, the terrain drops precipitously from enormous mountains to long stretches of semi-arid steppe. This bronze-colored scrubland with an occasional wind-twisted tree is traveled by long, straight gravel roads that wilt in distant mirage. Tierra del Fuego, "Land of Fire," was christened by Magellan when he saw smoke drifting from the fires of Fuegian Indians—the Ona, Haush, Yaghan, and Alacaluf—who had been living there for thousands of years. When Charles Darwin sailed through here on the *Beagle* in 1836, three hundred years after Magellan, he wrote, "No one can stand unmoved in these solitudes, without feeling that there is more in man than the mere breath of his body."

As I rest my forearms on the cap rail of the *Patagonia*, the ferry that today shuttles across First Narrows, I gaze on a scene almost unchanged since Magellan's voyage. The spit he christened Cape of the Eleven Thousand Virgins lies about twenty miles east, since renamed Cape Dungeness. To the west, the straits widen, then pinch again at Second Narrows before opening into a large basin near the town of Punta Arenas, Chile. Farther west, the chiseled faces of the Andes loom above the Pacific. There, rainfall is heavy and tidewater glaciers shift and moan, calving apartment-building-sized chunks of ice into the sea. *Williwaws*, frigid air that builds and releases suddenly from mountaintops, rush down valleys at more than 110 miles per hour, shaking trees and tearing at the water's surface. Frequent fog, heavy seas, and treacherous currents can make boat travel impossible through the fjords that stretch 1,200 miles up the Chilean coast.

On a clear and windless April day—the southern hemisphere's fall—I stand on the ferry's deck and hear the murmur of tidal currents plying the pebbled shore and erupting in boils that surge from below. Cormorants and penguins dive for fish. "I like to watch the way animals use tidal currents," says Gareth Davies, who leans his thick arms on the rail beside me. As founder and director of Aquatera, a consulting company based in the Orkney Islands of northern Scotland, Davies travels the world helping communities, corporations, and governments understand marine energy. We are both here for a tide and wave energy workshop in Punta Arenas, sponsored by Aquatera,

Chile's Ministry of Energy, and Alakaluf, an organization run by a local oceanographer, Sergio Andrade. But for the moment it is just Davies and me, watching the tide.

"These places draw us toward them," Davies says, his large brown eyes studying the surface. "They're dangerous, but their power and vibrancy are mesmerizing."

"I see them as metaphors of change," I tell him. "As one tide is born, another is dying. It's in these choked narrows that we witness the immense influence of the moon."

"I like that," Davies says. "Since the moon drives tide energy, maybe we should be calling it 'moon energy'!"

〰

A year earlier I had knocked on Davies's office door in Stromness, a small Orkney port town with narrow cobbled streets and tall stone buildings. I was there visiting the European Marine Energy Center (EMEC), the world's largest tide and wave energy test site, established in 2003. Davies, a heavyset PhD marine biologist in his early fifties, moved to the Orkney Islands in 2000 to help set up EMEC. His consulting company, Aquatera, is independent of EMEC but housed in the same building. Serendipitously, he was in his office when I knocked, having just returned from Paris and on his way to Japan to consult about potential tide energy in Naruto Straits.

We found a park bench overlooking Hoy Sound and the entrance to Scapa Flow, location of one of the test sites. The Orkney Islands were selected as Scotland's premier test site because of the fast currents, which are created in much the same way as the currents in the Straits of Magellan—in this case, the uneasy meeting of the North Sea and the Atlantic. As one body rises, water flows through the seventy Orkney Islands and skerries (rocky islets) to fill the neighboring ocean, bringing to life hundreds of high-energy passages. To the south, on the other side of Hoy, the infamous Pentland Firth separates the Orkney Islands from mainland Scotland. In that nine-mile-wide channel, the tide runs at ten to twelve knots, and sites are already leased for the long-term home of machines that pass testing.

"Most oceanographers agree that there are about 3.5 terawatts of raw power in the ocean's tides," said Davies. "That's equivalent—at any given moment—to about 3,500 large coal plants running at capacity." During my visit to the Orkney Islands, I learned that technological and environmental limitations make only a quarter of that raw tide energy harvestable. On a global scale, that probably isn't quite the silver bullet to end fossil fuel addiction, but it might be enough to play a meaningful role with other renewables, such as wind and solar. For communities and smaller countries surrounded by water, it could be transformative.

〜

Ten months later, Davies invited me to join him at Chile's first international marine energy workshop. It was to be a gathering of experts from Chile, Norway, Scotland, Canada, and the United States to discuss Chile's future in tide and wave energy. Having read about Chile's commitment to explore tide energy, I instantly agreed; later I met Davies and his assistant, Tom Wills, in Santiago so we could drive down the coast together to the workshop.

As Davies and I study the water's churned and dimpled surface from our perch on the First Narrows ferry, he tells me of Scotland's marine energy history and the lessons it might offer Chile. In 1982 the Scottish government issued a mandate to generate 100 percent of its electricity with renewable resources by 2020—by far the world's most ambitious renewable energy target. For comparison, the European Union aims for 20 percent by 2020, and Germany, with some of the world's most aggressive renewable standards, aims for 30 percent by 2020 and 80 percent by 2050. Chile's goal is 10 percent by 2024. The United States has *no* federal standard.

The Orkney Islands were chosen as the Scottish center for renewables because of the exceptional wind, wave, and tidal resources as well as its active marine industry. EMEC was set up (largely with European Union and also some Scottish government funds) to offer test sites for international developers. To kick-start the testing process, the government initiated in 2010 the Saltire Prize, a £10 million purse for the first device to generate 100 gigawatt hours "using only the power of the sea" for a continuous two-year period. Since 2003, EMEC has overseen the testing of more than ten different tide machines, most of them turbines that look and function like underwater windmills. To date, none of them has come close to winning the Saltire Prize.

"Scotland certainly doesn't have all the answers," says Davies, "but we've been at it longer than anyone else. Chile's on the other end of the spectrum—just beginning. They have great marine resources. And right now they have the advantage of studying what's happening in other countries and choosing their own path into this arena."

When it comes to renewable energy, Chile's noodlelike shape may be one of its best assets. Squeezed between the Pacific and the Andes, the country stretches through sixty degrees of latitude, capturing almost every kind of terrain and bioregion, from deserts to rainforests, mountains to conifer forests, beaches to rocky shorelines. The 41,000-square-mile Atacama Desert in the far north competes for the best solar potential anywhere on the planet. Chile's 4,000 miles of Pacific coastline facing due west, battered by wind and swell from southern hemisphere storms, has almost limitless offshore wind and wave energy potential. So it is with geothermal energy too: balancing atop the Pacific Ring of Fire, this country is rocked by an average of one and a half earthquakes a year, magnitude five or larger. And the tides and sinewy fjords of southern Chile create thousands of miles of high-energy tidal passages.

"Harnessing even a fraction of the country's renewable energy could easily meet all of Chile's electricity needs, and more," says Davies as we leave the First Narrows ferry and head toward Punta Arenas.

≈

On the morning of the workshop, the convivial chatter of sixty people hushes as Don Carlos Barria from the Chilean Ministry of Energy
begins:

> Bienvenidos a todos. Gracias por venir desde largas distancias para estar
> aquí para ayudar a Chile navegar su camino en el ámbito de la energía
> marina. Estamos entusiasmados por las discusiones que tendrán lugar en
> los próximos dos días.

"Welcome, everyone," he says. "Thank you for coming from so far
away to help Chile navigate its way in the marine energy field. We're
excited about the discussions that will take place in the next two
days." The Chilean government, he explains, recognizes its renewable
resources and is developing a national energy strategy that looks to
the year 2030. He points out that although Chile is one of the richest
countries in South America, it's among the poorest when it comes to
energy, importing at great expense more than 75 percent of its fuel in
the form of natural gas, coal, and oil. "We need to develop local renewable energy for many reasons," he says, "but it's not as simple as just
buying and installing the right equipment. We must also rewrite our
laws and create new supporting infrastructure."

〜

From a window seat in the hotel conference room, I can see the deep
blue-green of the straits and the distant snowy mountains of Tierra del
Fuego. My mind wanders back to Magellan, who cleverly found a way
to use the tide and wind to claw through the straits. An expedition like
his—the first to circumnavigate and prove the world was round like an
apple, not flat like a table—was fueled entirely by wind and will. But
the wind that filled his sails hundreds of years ago pushed more than
ships. It was also driving thousands of mills across Europe and Asia, as
it had been for centuries. As early as the tenth century, windmills were
used in Persia to pump water for irrigation and in Europe to grind
corn. The technology reached a peak in the Netherlands between 1500

and 1650, where windmills were put to use draining agricultural fields and grinding wheat, spices, and fiber to make cloth.

The tide energy that cursed Magellan's navigation of the Dragon's Tail in 1520 had already been recognized as a blessing in other parts of the world. Historians believe the first tide mill may have appeared in Roman times (200–400 CE). Confirmed evidence of a mill built on Strangford Lough in northern Ireland dates back to the late eighth century. By the seventeenth century, the simple elegance of tide mill technology had reached a tipping point, and more than a thousand were operating on the coasts of the United Kingdom, the Netherlands, Portugal, Belgium, France, and Spain. The Tagus estuary in Portugal, south of Lisbon, had as many as forty during this period. Like windmills, they were put to work milling cereals, but also stamping copper, crushing or rolling kaolin (clay) and tree bark, pumping brine for extracting salt, preparing pulp for paper, crushing ice, and sawing wood. It's not unlikely that a tide mill on the south coast of Spain provided the flour used by Magellan to make hardtack, a twice-cooked brittle biscuit that was the expedition's main food supply.

Tide mill engineering was straightforward. A house with a wooden paddlewheel was built across the tidal estuary. As the tide rose, gates were opened to fill a holding pond, or millpond, on the upland side. When the tide peaked, the gates were closed, trapping the water. The gates stayed closed as the tide in the estuary dropped, creating a difference in water level (a hydraulic head). When the tide on the seaward side had dropped sufficiently—in this case three or four feet—a jet of water was released from the millpond through a chute, or sluice gate, aimed at the waterwheel's wooden paddles. The wheel would rotate, never very fast, driving gears with iron or wooden cogs, which would eventually turn the grinding stones. On a large tide, the wheel would run five or six hours.

This technology was introduced to North America's eastern seaboard in the early seventeenth century and quickly spread from Boston to the Bay of Fundy. The first tide mill on record was built in Port Royal,

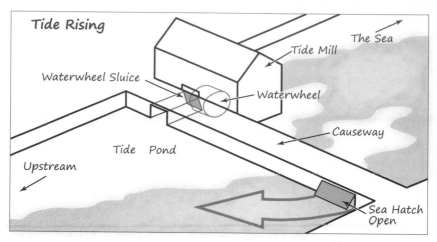

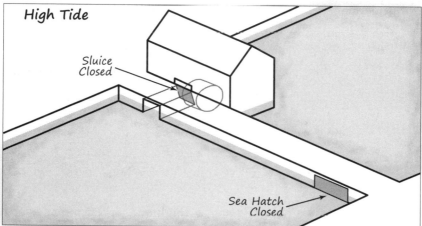

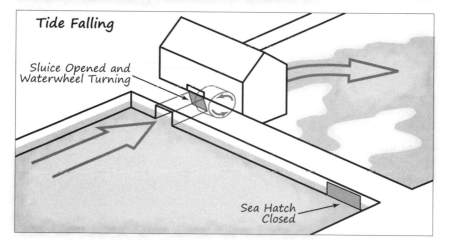

The workings of a tide mill.

Nova Scotia, in 1607. A hundred or more were operating in the late eighteenth century. These mills were located in estuaries with large tides and convenient access for oceangoing ships, enabling the import of raw materials and the export of mill-processed goods, including lumber and cloth.

Tide mill communities flourished in Europe and North America. Millers worked on the tide's schedule, an hour later each day and peaking during the new and full moon. When the heavy tide gates closed with a resounding clunk, whether night or day, all the community knew it was high tide and the millpond was full: it was time to get ready for work. The waterwheel would start turning in a few hours.

〜

In 2013 I took a two-hour train ride from London to see the Eling Tide Mill, which has straddled the mouth of the Bartley River near Southampton for 230 years. From the train station, a cabdriver shuttled me through progressively smaller villages and narrower streets until he stopped in front of an unassuming red-brick building. I stood on the street for a few minutes after the cab left, taking in the building and the estuary beyond. The tide was on its way up, slipping under fifty or more boats crowded into the anchorage. Many were still aground and listing in the mud, their masts leaning every which way. I imagined sailing ships laden with raw corn and wheat standing off the estuary during the mill's heyday in the eighteenth and nineteenth centuries, waiting for enough water to float them in. At high slack, they would have sailed in, tied off to the mill just long enough to exchange raw goods for flour and set sail again before getting caught by the falling tide. Raw goods were also brought from inland by horse and carriage across the one-lane bridge that still connects the mill with the rest of Eling.

The mill itself has three floors and a steeply pitched slate roof. It was built in 1785, but there's evidence that the site may have supported two corn mills, one of them fueled by the tide, as early as 1086. The present building, damaged and restored several times, is now under the care of the town council, which manages a museum in the old miller's quarters as well as the operation of the mill itself, which today produces a monthly average of 1,700 pounds of whole-meal brown and strong white flour. Some of this is baked into breads and cookies sold in the museum gift shop, and the balance is sold to local bakeries. The three-pound bags for sale at the museum boast "Canute 100% Stoneground: Milled from English bread-making grain for a fuller flavour at the only working tide mill in the United Kingdom." On the back is a recipe for a "delicious 2 lb loaf."

The Canute packaging is a reference to King Canute of England, Denmark, and Norway, 995–1035 CE. Legend has it that to dispel his court's claim that he was powerful enough to "command the tides of the sea to go back," Canute directed his throne to be carried to the seashore—ostensibly near Southampton—and sat as the tide came in, commanding the water to "advance no farther." When the tide paid no heed, the king's point was made: sovereign power might be great, but nothing was greater than the hand of God. (Indeed, during Canute's day the cause of the tide was still a complete mystery.)

Today the lead miller is David Plunkett, a retired stonemason with deep-set blue eyes and ruddy cheeks. I had missed the previous tide cycle, and the next wouldn't start until six o'clock that evening, after the building was closed. Plunkett, however, invited me to stay. He and his apprentice, Andrew Turpin, would run the mill, and I would be put to work lifting sacks of grain and refilling the grist bin.

After returning from a dinner of stew and beer at the local pub, Plunkett flicked the tide gauge with his thumb. "We'll be ready to fire up the wheel in about fifteen minutes," he said as he pumped grease into the main waterwheel's brass cap and donned a white apron. Turpin grabbed a bag of raw wheat supplied by a local farm and asked me to pour two scoopfuls in the grist bin. The two of them walked through the building, methodically adjusting valves, cables, and the two four-foot-diameter grinding stones, each weighing a ton, fabricated in France of composite granite. The massive timber-framed structure, some of it cobbled together or partially eaten by insects, reminded me of an old ship. As if below decks, I had to duck under low-slung oak beams and knee braces.

"Here we go," announced Plunkett as he turned an iron valve opening the sluice gates. Laboring at first, the wheel eventually rumbled to speed. The whole place seemed to come alive, creaking and groaning like a ship at sea. Windows shook. Iron latches and levers rattled. A spoon and pencil chattered in a jam jar. Upstairs, the grist bin's *tick tick tick* confirmed that grain was being fed to the grinding stones. When I looked up from the swishing and gurgling waterwheel, a stream of golden flour was pouring from the chute.

The waterwheel ran for four and a half hours. Through the windows, I watched the tide disappear from the harbor, leaving all the boats aground. The mill's wheelwash gushed into the estuary like a whitewater rapid. Plunkett and Turpin were constantly tinkering with valves, watching, touching, listening. A fine powder had filled the air as soon as we started producing flour, settling on everything, including our eyebrows and lashes. The place smelled like baking bread, hot grease, and low tide.

A few times an hour, Plunkett put his hand under the warm flour spilling from the chute. Spreading it on his open palm, he'd run his thumb through it, feeling for temperature and texture. "What we're after is a balance of wheel speed and distance between the two grinding stones," he said. "If the stones are too close, the flour gets hot and sticky. If they're too far apart or we're turning too slow, the flour's coarse."

"How do you know when all the elements are perfectly tuned?" I asked.

"There's a vibration that feels just right," he answered. "I can feel the humming in my skull."

The waterwheel shut down when the rising tide reached its axle and there was too much drag for it to continue turning. I drank a celebratory beer with Plunkett and Turpin at the local pub and caught a midnight train back to London, carrying a bag of fresh Canute Stoneground, flour of the tide. The train's soothing hum reminded me of the tide mill and Plunkett's words: "After fifteen years, I'm still amazed," he had said. "There are no engines, no gasoline, no coal, no electricity except for a couple of light circuits. All this is powered by the moon and tides."

～

The energy sources of fire, petroleum, coal, and their by-products have been used by humans for a long time—fire for 400,000 years, petroleum for 4,000 years, and coal for at least 3,000 years.

Petroleum was used by the Babylonians to make asphalt as early as 1894 BCE. Almost four thousand years later, Londoners were distilling it into kerosene to fuel streetlamps (replacing whale oil). Natural gas, a by-product of coal mining and oil drilling, was also used for lamps.

Coal, first burned in ceremonies by the ancient Romans, was used by London blacksmiths in the twelfth century. Its smelly, dirty properties were well known. When Queen Eleanor visited Nottingham in 1257, the air was so sooty she fled, fearing for her health. Forty years later, coal burning was banned because of its polluting qualities.

By the time Magellan set sail from Spain in the sixteenth century, Great Britain was facing perhaps the world's first energy crisis. Wood, the fuel of choice for heating and cooking, was in low supply. Forests were shrinking.

Coal, dirty or not, was the ready answer. "By the mid-1600s," writes Barbara Freese in *Coal: A Human History*, "Londoners did not merely welcome coal into their homes, they were desperate to have it." Mined at nearby Newcastle and Tyne River, coal solved London's energy crisis but at the expense of stifling air pollution. In 1700, addressing the insidious smoke that plagued London, essayist Timothy Nourse wrote of London, "There is not a more nasty and a more unpleasant Place."

That was more than three hundred years ago.

The steam engine, invented in 1781 and fueled by coal, catapulted Britain and eventually the rest of the world into the Industrial Revolution. By the end of the nineteenth century, steam-fueled mills had largely displaced wind- and tide mills. The Eling tide mill itself was converted to steam in 1890. Thirty years later, unable to compete in the flour and textile industries, Eling was producing only animal feed.

Prior to the Industrial Revolution, the world's consumption of fossil fuels was negligible. By the year 2000, our consumption had grown by a factor of forty, and by 2015 it had grown by a factor of eighty. In the overall global energy budget, 50 percent comes from coal, 20 percent from oil, 15 percent from natural gas, 5 percent from nuclear fission, and 10 percent from renewable sources such as biomass (wood burning, etc.), hydropower, wind, and the sun.

Accelerated fossil fuel use is tied to the planet's population growth, and such use is tied to lethal concentrations of pollutants that have changed the planet's chemistry and lessened the quality of life for everything that lives on it. A growing concern about this problem was best signaled by the landmark Paris climate deal in December 2015. But the awareness has been growing for years. In an August 2013 letter to the *New York Times,* four former Republican administrators of the Environmental Protection Agency (EPA) wrote, "There is no longer any credible scientific debate about the basic facts: our world continues to warm, with the last decade the hottest in modern records, and the deep ocean warming faster than the earth's atmosphere. Sea level is rising. Arctic Sea ice is melting years faster than projected." The letter calls for a reduction in carbon dioxide emissions and increased investment in clean, renewable energy. Signed by William D. Ruckelshaus, Lee M. Thomas, William K. Reilly, and Christine Todd Whitman (representing forty-three years of EPA leadership), the letter concludes: "The only uncertainty about our warming world is how bad the changes will get, and how soon. What is most clear is that there is no time to waste."

≈

From our Punta Arenas hotel conference room, the view of Magellan Straits is strikingly raw and elemental, offering no hint of an environmentally troubled world. How did we lose our way so quickly? What level of environmental health will we use as a baseline for recovery, if there is to be a recovery? Gareth Davies, who takes the podium next, begins as if his words were the next sentence of the *New York Times* letter: "I believe the world is in a hurry for us to come up with a better solution for energy," he says, "because we're not in a good place at the moment. The energy we use is killing the world.

"Only 10 percent of the world's fuel supply currently comes from clean, renewable energy," Davies explains. "And of that, only 2 percent comes from tidal energy. But these numbers are growing. In the next twenty years, I hope to see the renewable sector increase to 40 percent." According to Davies and many of his colleagues, tide energy will probably remain a small but important part of that 40 percent. This is not because harnessing the tide is so difficult, but because in terms of raw available power, wind and solar far outstrip all the other fuels combined, renewable or not.

What tide energy *does* offer is predictability—more than most other renewables. The wind doesn't always blow and the sun is sometimes hidden by clouds, but the tide *always* ebbs and flows.

"For tide energy to flourish," Davies continues, "we will have to resolve many of the same issues we face with other renewables. Getting the engineering right—building devices that work—is just one of them. Launching and servicing the devices is another, and so is political support, both nationally and internationally. We need more mechanisms in place that provide financial support during this early—and expensive—development stage. Utility companies, which frequently have near-monopolies on supplying power, will have to change. Environmental concerns need to be addressed, and the communities that will live with these devices in their backyard have to be on board. For them it means jobs and cheaper, cleaner energy. But it also means making reasonable compromises for the sake of the big picture."

From my time with Davies as well as my own research, I've learned to appreciate his cautious optimism. Technology is one of the obstacles for all renewable energy sources. Wind and solar technologies are far more advanced than marine technologies, still in their nascent stages. Designs similar to the Eling Tide Mill, a barrage across a tidal estuary, have been revisited several times in the past hundred years, often in response to a fossil fuel crisis. Perhaps the earliest modern-day effort was in Passamaquoddy Bay on the border of Maine and New Brunswick, which has a twenty-six-foot tide and a network of inlets that could be blocked off by barrages to form inland "pools." The hydraulic head of these pools could then be utilized to funnel water through electricity-generating turbines.

≈

The U.S. government first studied Passamaquoddy in the 1920s but determined the engineering was too ambitious and costly. The effort was renewed a decade later during the Depression's public works program, only to fizzle in the face of opposition from utility companies. However, with an estimated capacity of a thousand megawatts (enough to supply more than a million homes), the project has never really died. Over the next twenty years, various engineering schemes were proposed, some of which were joint ventures with Canada. An international commission report, released in 1961, concluded: "It is evident that construction of the tidal power project by itself is economically unfeasible by a wide margin. . . . Either alone or in combination with auxiliary sources, [the tidal project] would not permit power to be produced at a price which is competitive with the price of power from alternative sources."

President John F. Kennedy was not convinced of the commission's findings and asked for a review by the Department of the Interior. The department's report recommended that the project proceed. "I am pleased to meet today," said Kennedy from the flower gardens of the White House on July 16, 1963, "to discuss the report on the

International Passamaquoddy Tidal Project submitted by Secretary Udall.... The report reveals that this unique international power complex can provide American and Canadian markets with over a million kilowatts of firm power." Kennedy explained that New England's electricity rates were among the highest in the United States, and the Passamaquoddy project could reduce those rates by 25 percent. "Harnessing the energy of the tides," he continued, "is an exciting technological undertaking.... Each day, over a million kilowatts of power surge in and out of the Passamaquoddy Bay. Man needs only to exercise his engineering ingenuity to convert the ocean's surge into a great national asset."

Private utility interests again mobilized in response to Kennedy's plan. A team of consulting engineers was hired to find flaws in the Department of the Interior report. Their rebuttal convinced the public that the project was too expensive, time-consuming, and loaded with engineering pitfalls. Between these criticisms and the emerging interest in nuclear power—at the time considered the next best and cheapest solution to high fuel costs—Passamaquoddy foundered. Kennedy's bold advocacy of tide energy ended abruptly with his assassination in November 1963, just four months after his groundbreaking speech at the White House. His words are inscribed above EMEC's door: "The problems of the world cannot possibly be solved by skeptics or cynics whose horizons are limited by the obvious realities. We need people who can dream of things that never were."

In the twenty years that followed, barrage-type tide energy projects were completed in France, Russia, and Canada. The 240-megawatt French barrage was built in 1966 across the La Rance estuary near Saint-Malo in Brittany; the 1.7-megawatt Russian barrage was built in 1968 at Kislaya Guba in northeastern Siberia; the 20-megawatt Canadian barrage was built in 1984 in Bay of Fundy's Annapolis Royal. These plants are still in operation today. Canada considered several more installations in the Bay of Fundy during the Arab oil crisis of the mid-1970s, but the plans were abandoned when scientists like Peter Hicklin and Chis Garrett showed that they would have adverse affects on the tidal and biological systems.

The good news about barrage-type tide energy projects is they generate large amounts of electricity (at 240 megawatts, La Rance produces two hundred times more electricity than any single existing nonbarrage tide project). But barrages have fallen from favor due to their astronomical building costs and negative effects on the environment. Because they function like a dam, they block fish migrations and interfere with the natural tide exchange between land and sea, causing severe upland silting, which suffocates vibrant estuarine ecology.

Tide barrages have not gone away, however. In 2011 Korea opened the 254-megawatt Sihwa Lake Power Station, overtaking La Rance as the world's largest. Proposals for a barrage across the Severn River, which separates England and Wales, have been considered since the nineteenth century; in 2013 the latest proposal was declined because, according to the review committee, the developer had failed "to answer serious environmental and economic concerns." A 250-megawatt tidal energy lagoon, a modified barrage designed to generate electricity during both incoming and outgoing tides, is also proposed for Swansea in South Wales.

≈

The approach to tide energy has changed in the last thirty years. The newest efforts have moved away from expensive and ecologically disruptive barrages toward devices that sit below the surface in a tide channel and harvest energy from the incoming and outgoing currents. This approach is called *in-stream*, a term that applies to both tidal and river installations, although tidal is far more developed. Most of these in-stream devices look like an underwater windmill, turning in the current the way a windmill turns in the wind. But some sit on the surface and take advantage of the tide's vertical rise and fall, and a few are designed to soar in the water column like a kite or a bird. The electricity these devices generate is transmitted to shore by heavy cables and from there to a nearby power grid.

Generally, in-stream devices, or machines, come in three sizes: commercial, community, and micro. Commercial-scale devices are huge—with blades up to seventy feet in diameter—and are designed to generate as much electricity as possible: five hundred kilowatts and up. Once these commercial units achieve a certain level of success, they'll be installed in arrays on the ocean bottom, like a wind turbine farm.

When I was in the Orkney Islands, I stumbled upon a couple of these devices being repaired on a commercial dock near the town of Kirkwall. One, owned by Rolls Royce, is a bulbous orange thing about the size of a very large recreational vehicle, with three fat propellers sticking out one end. When launched, the machine is towed on a barge and positioned above a turbulent tidal channel. During the ten or twenty minutes of slack water, the machine is lowered by crane through the water column before touching down on its seabed cradle. All this must be done with the precision of landing a space module. If something goes wrong—the weather turns or a widget malfunctions—the mission is aborted, at huge expense, and everyone goes home until the next slack tide.

On the dock for repair, the Rolls Royce machine sat awkwardly on chocks high in the air while welders and engineers worked day and night to get it back in the water. It was fenced in like a large zoo animal, with worklights mounted on steel scaffolding and arm-thick cables strewn on the ground. The crew kept mum when I asked questions, though I did learn that another machine, twice the size of this one, was under construction outside London.

Down the dock, another device was tied alongside, this one owned by Orkney-based Scotrenewables. It too is tubular, but 120 feet long and slender like a submarine. Its yellow paint looked fresh. It has folding "wings," each with a two-bladed turbine that tucks in close while the unit is towed from site to site, or from site to dock for repairs.

Unlike many of the devices being tested, this one stays on the surface and unfolds its wings into the tidal currents below.

Devices are in and out of the water regularly, being repaired, tuned, and tweaked. Many are tested at EMEC in the Orkney Islands, but some are tested independently or at the Fundy Ocean Research Centre for Energy (FORCE) in Canada's Bay of Fundy. Like EMEC, FORCE is an international testing site funded largely by the government. Located in Minas Basin near the town of Parrsboro, Nova Scotia, it has four berths to lease. The first device, designed by OpenHydro, was installed in 2008 and pulled a year later for repairs. Minas Basin's strong currents make this site one of the more challenging test locations.

The commercial device with the longest continual in-water operation is SeaGen, installed in 2008 at Strangford Lough, Northern Ireland. (Strangford Lough was also the site of the eighth-century Nendrum tide mill, the oldest on record.) SeaGen, built by Marine Current Turbines, has a center tower drilled into the seabed that rises thirty feet above the ocean's surface. Two swiveling arms, each with twin fifty-four-foot blades, extend below the surface. When rotating at capacity,

The SeaGen in Strangford Lough.

A sampling of tide machines, all under development.

the SeaGen can generate 1.2 megawatts of electricity, enough for 1,500 homes.

While there's a strong financial push to develop commercial-size machines, midsize devices are also being developed for remote communities and coastal aquaculture. New York–based Verdant has had community-scale tide machines in the East River since 2002 and is planning to install more in the future. Smaller devices—known as microscale—are designed to generate just a few watts, enough for a navigation light or an oceanographic sensor. Some are so small they flutter in the tidal stream like a fishing lure.

No matter the size or approach, these devices must hold steadfast in almost impossible conditions—where Magellan and no other right-minded sailor would want to linger. First, they must endure the six-hour storm of an incoming tide, with heavy currents corkscrewing in every direction. Then they must withstand the same storm from the opposite direction—the outgoing tide—for another six hours, all while dependably producing and delivering electricity to shore.

Every step of this process is under development, from blades to gaskets, from hinges to couplers to cable stout enough to carry the electrical loads. And, as if that's not enough, there are the challenges (and frustrations) of obtaining licensing from regulatory agencies, negotiating with reluctant utility companies, securing permits and offshore leases, weighing environmental impacts, and evaluating potential interference with established uses such as fishing, shipping, coastal security, and recreation.

≈

"Sometimes I get so discouraged I want to quit," says Chris Sauer of Maine's Ocean Renewable Power Company (ORPC), whose 150-kilowatt machine, TidGen, recently completed a continuous one-and-a-half-year in-water, grid-tied operation in Eastport, Maine, the first of its kind in the United States.

Developers are generally skittish about divulging project information, but over breakfast on the first day of the Punta Arenas workshop, Sauer shares the details of ORPC's history. He will give a presentation for the group later today, but at the hotel café he is surprisingly candid about the story behind the story. What he tells me is about one company and the evolution of one machine, but it's representative of the experience of every developer, big or small.

Sauer, in his late sixties, is tall and thin. His full head of frizzy gray hair accentuates his zeal, especially for marine renewables. With a background in engineering and business management, he spent forty years in the coal, oil, and natural gas industry. He was retired and living in Florida when in 2004 he was asked to consult about generating power from the Florida Current. "I knew nothing about the ocean," he says. "But I thought, if we could use the ocean to generate electricity, that would be something!"

A couple of years later, Sauer took over ORPC and submitted his first design concept to the Navy for review. "They hated it," he says. "Among a long list of things, they said the blades were too big to withstand the stresses." With their help, Sauer designed a new machine using cross-flow turbines and a high-torque, low-rpm generator, all to minimize stress. A scaled-down version was built and tested in the summer of 2007. "It barely worked," says Sauer, "but we got enough data to know the basic turbine concept was sound." His team also realized that cross-flow turbines, which look and operate like a manual lawn mower, rotate the same direction no matter which way the current is flowing. "So, *bingo*," Sauer says, "we can do tidal!"

Ocean Renewable Power Company's TidGen, designed to rotate on both incoming and outgoing tides.

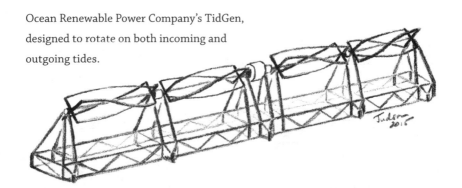

From that point, ORPC's focus shifted from offshore currents to in-stream tidal and river. The company applied to the U.S. Federal Energy Regulatory Commission (FERC) for permits to launch machines in Cobscook Bay in Eastport, Maine, and Cook Inlet in Alaska. "I had been to Alaska during my days of working with fossil fuels and knew how strong the tidal currents were. Other than the Bay of Fundy, Cook Inlet is probably the most energetic site in North America."

The next machine was tested for a full year in Cobscook Bay. It generated electricity, but it was too scrambled to be grid-compatible. "You've got to be able to plug your hair dryer into these things," says Sauer. An improved machine went in the water in 2010 and came out a year and a half later. This one was fraught with problems too—finicky electronics, failed connectors, alignment issues—but it succeeded in delivering grid-compatible electricity.

In 2012 ORPC got its pilot-project license for an installation in Cobscook Bay, the second issued in the United States for tide energy but the first actually built. "It was an enormously time-consuming and expensive process," says Sauer. "FERC deals with inland dams, so tidal energy was completely new to them. They're good people, but this was so new they didn't even know what questions to ask." ORPC launched the TidGen that summer in Cobscook Bay and left it in for a year, where it supplied enough power for about thirty homes.

Encouraged, Sauer began looking at the international market: Canada, the United Kingdom, France, Japan, Taiwan, China, Korea. All these countries have coastal zones with large tides and currents, and all are exploring tide energy, along with Australia, New Zealand, Norway, and Russia, among others. "These places are all vastly different," says Sauer. "In each case, you have to ask a lot of questions: How much tidal energy is there? How accessible is it? Are there supporting marine facilities? What are the environmental concerns? Is there an electrical need, or *load*, nearby? What's the political climate? Are there economic incentives? Finally, you have to ask if they like Americans, because not every country does."

As we finish breakfast, Sauer tells me he's drawn to Chile because it's got great tidal resources and it's friendly, politically stable, and market-oriented. "And," he says, "I don't see another tide company knocking them dead down here."

When I ask which sites are likely to be developed first in Chile, he explains that it probably won't be First Narrows in Magellan Straits. "It's a good site, but too far from the grid," he says, "and even if it could be grid-tied to Punta Arenas [the closest town], the loads here are not demanding enough since they use natural gas and not electricity to heat their homes." Sauer believes the best site is Chacao Channel near Puerto Montt, about 1,500 miles north of Punta Arenas. "I don't know if I've seen a better tide energy site than Chacao—anywhere," he says.

~

Gareth Davies and his assistant Tom Wills took me to see Chacao Channel. We met in Santiago a few days before the conference and drove down the coast, stopping at Valparaíso, Concepción, and Valdivia where Wills and Davies gave presentations at universities and met with engineers and oceanographers to explore partnerships for Aquatera's work in Chile.

As we drove into Puerto Montt, the closest city to Chacao Channel and the region's commercial and industrial hub, I studied the landscape out my backseat window. Billboards flicked by advertising Husqvarna chain saws, fertilizer, Mack trucks, and Thule luggage racks. Fields were plowed under for winter. The ever-present Andes loomed in the east, and the snow-topped volcanoes of Osorno, Puntiagudo, Calbuco, and Hornopiren towered to the south and west. As we got closer to the port, signs advertising escape routes in the event of earthquake, tsunami, or volcano eruption were everywhere. On the wharf, a café owner who served us a lunch of fish soup and empanadas told us that several years ago all three warnings sounded at once. "Where do you run to, then?" she asked.

From Puerto Montt, we drove west toward the open sea and arrived at Chacao Channel by late afternoon, in time to catch a small ferry that crosses the channel to Chiloé Island. The day was overcast, but I could see in all directions. The mustard-colored bluffs and brilliant green fields of Chiloé, a farming and fishing community, are only a few miles on the other side of the channel. Toward the west, the gray-blue ocean spilled into the sky. The volcanoes I had seen earlier were still prominent. The smells of diesel and fish in Puerto Montt's inner harbor gradually gave way to aromas of heavy salt air and evergreens.

Davies wasted no time making his way to the boat's bow. He noted a seal or two, a grebe, a few white-breasted cormorants, and a flock of gulls. The current was ebbing, like a large, deep river lumbering implacably toward the sea. "I've wanted to see this place for a long time," Davies said. What makes this site so perfect, he explained, is that it's in a rural area yet close to an industrial port with a large electrical load. (Puerto Montt's power grid connects to Santiago and Valparaíso, where about 50 percent of Chile's population lives.) The channel itself is long and wide, 150 feet deep, fairly flat-bottomed with few large reefs or obstructions. It's protected from the open ocean and, the salient point, has a twenty-one-foot tide with five to six knots of "orderly" (meaning not too turbulent) tidal current. When I asked how much energy was in the channel, Davies replied that tide models showed about 600–800 megawatts. "Not all that is extractable, though," said Davies.

Scientists agree that 3.5 terawatts of global tide energy is the correct *theoretical* measure of the total, based on the physics. The electricity that a single site like Chacao Channel can actually generate is much less, given the technical, social, economic, and environmental realities. "The best sites in the world usually have between 500 and 1,000 megawatts of theoretical energy," explained Davies. "But from a practical view, we can capture only about 20 to 30 percent of that. It's this practical view that everyone cares about. That's the energy that can actually be delivered to your home, but it's not easy to estimate that in advance of actually having a device in the water because there are so many technical and socioeconomic unknowns."

〜

The conference in Punta Arenas is Chile's first attempt to get a handle on these practical issues. The government is well aware of the country's extraordinary tidal resources but also aware of the realities of exploring uncharted territory. Addressing socioeconomic unknowns, for example, is one of the issues that must be confronted by project development, which is a process parallel to—yet very different from—device development. Device development is all about building a workable machine; project development is all about public process and securing the permits necessary to put the machines in the water. Most device developers want to focus only on engineering issues. They're not usually trained nor do they have the resources to pursue onerous public and regulatory processes, such as seabed licensing and environmental risk assessments. This is where the test sites of EMEC and FORCE, as well as a few smaller test sites, come into play. At these sites, much of the project development is already done—public process completed, licensing secured, monitoring equipment in place, electrical cables installed, and so forth. Once an application for a test berth is accepted, a developer is free to undertake the challenges—and they're formidable enough—of building and launching a workable device.

What these sites offer is not trivial. Good project development, especially in territory where the laws are unclear or nonexistent, is time-consuming and expensive. It took years (and millions of subsidized dollars) to get EMEC and FORCE up and running. To be successful, project developers need teams of scientists, lawyers, community leaders, engineers, and financiers. They also need people who are extraordinarily skilled at listening to and understanding the concerns raised by people who live and work near a proposed installation. A local biologist, for instance, might be concerned about the effects of underwater noise, electromagnetic fields, or potential collisions between rotating blades and marine mammals. A politician might focus on new jobs. Property owners may fuss loudly about visual impacts. A tribal leader might worry that a submerged device will interfere with the tribe's "usual and accustomed" fishing rights.

"We're all inventing this process together," said Davies as we drove back to Puerto Montt, where we caught a plane down to Punta Arenas. "In Orkney, we had lots of public hearings and discussions with the community. Most people were on board right away, but there was a small group of stakeholders—environmentalists, fishers, recreational outfitters—who were worried about possible impacts. As a compromise, we agreed to install two machines and monitor them for three years. That test showed no negative impacts. Today, almost fifteen years after EMEC was established, the community is completely on board. More than that. They're proud of what they're doing. Over 300 people are employed in Orkney's marine energy field, and a lot of the locals have invested considerable money in the projects."

The project development process is the same everywhere in the world, although particulars vary. A site like Naruto Narrows in Japan has excellent tide energy, but it's smack in the middle of a commercial tuna fishery. Many of Canada's best west coast sites are in remote and unpopulated regions with small or no power load. Canada's east coast, and specifically the Bay of Fundy, is what many call "the Everest of tidal energy." But the claim is also its nemesis. Whereas five- to seven-knot currents are usually considered optimal, the FORCE test site in Fundy's Minas Basin has an average of twelve knots, with peaks of more than sixteen knots. "That will rip a machine to shreds," said Davies. And it has. The first device tested there, a sixty-foot-diameter OpenHydro turbine installed in 2006, lost its blades in the first three weeks. "The bad news about that," a FORCE representative told me, "is we learned these machines don't yet have what it takes to withstand Fundy's turbulence. The good news is there's more power down there than we thought."

The Orkney Islands also have their challenges. They have tremendous tide energy resources, but the power cables connecting

Orkney to mainland Scotland are already at capacity. They require a multibillion-dollar upgrade to justify permanent tide energy development in the islands. France has good tidal energy resources too, but the incentive for development is dampened by inexpensive nuclear-generated electricity (largely government subsidized). Government policies make all the difference in France and elsewhere, especially when research and development costs are high. Scotland, for example, has both "push" and "pull" mechanisms in place. "Push" mechanisms provide low-interest loans, grants, and other subsidies for start-up development. "Pull" mechanisms are market-driven and can take the form of tax credits. Scotland's Renewable Obligation Credit is one of the most progressive pull mechanisms because it provides a larger credit to the least developed technologies: wave and tide.

Chile has already provided grants—a push mechanism—for "water-ready" projects, which will get devices in the water and working. But for the technology to mature, pull mechanisms are essential. Getting those in place will be one of Chile's largest challenges because Chile's electrical utilities are privately owned. The government can advise on practices, but by constitutional law it cannot impose regulations, such as feed-in tariffs, renewable energy credits, and so forth.

〜

The United States has a basically supportive government and industry, but despite a few executive orders by President Obama, we have no renewable energy policy, which allows environmental regulations to trump development push. Our best potential sites are the San Francisco Bay; Puget Sound, Alaska; and New England.

A feasibility study commissioned by the City of San Francisco in 2006 determined the theoretical tidal power under the Golden Gate Bridge at 12–15 megawatts. The study concluded that only 10 percent of that was realistically harvestable. "The sensitivity of the Bay's ecological system," the report states, "limits the available extractable power from Golden Gate flows to between one and two megawatts."

The project has been shelved for now, as the cost of development far exceeds the value of the electricity it would produce.

Alaska's Cook Inlet, where Chris Sauer and ORPC have an interest, has exceptional tidal resources, but the downside is severe weather, water-born debris (such as trees), and, due to oil revenues, little or no political incentive for developing renewables.

In Washington State, Snohomish County Public Utility District No. 1 (the twelfth largest PUD in the country) planned to install two eighteen-foot-diameter, 150-kilowatt machines. Under FERC permits, they would have been in Puget Sound's Admiralty Inlet for three to five years, primarily to learn from the process, test the equipment, and gain additional information on environmental monitoring. This is an ideal site for many reasons: the average maximum is five knots; the depth is a consistent 180 feet; the bottom, composed of hard glacial clay, is fairly flat with few rocky outcroppings; and the load centers of Everett, Seattle, and Tacoma are close by.

After years of research, much of the political and technical preparations were in place, but due to funding shortages, the project was abandoned before the machines were installed.

Besides funding, the issue that may put a cold stop on tide energy development in Admiralty Inlet—and all of Puget Sound—is environmental. Three federally listed endangered species live there: salmon, rockfish, and the iconic killer whale. If the Snohomish test turbines had produced even a hint of adversity for one of these species, the machines would have been pulled.

"We left no stone unturned," said Andrea Copping when I spoke to her several months earlier. Copping, who works for a laboratory run by Batelle Memorial Institute Research Center in Seattle, was part of the team offering technical assistance to Snohomish PUD. "They worked on this for eight years, beginning with public meetings up and down Puget Sound, as well as with the Suquamish, Lummi, Duwamish, and Tulalip tribes." They evaluated seven different tide sites, applied for FERC licensing, and responded to concerns raised by U.S. Fish and Wildlife and the National Oceanographic and Atmospheric Administration (NOAA), the latter being the agency responsible for upholding the Endangered Species Act. "NOAA would ask questions," Copping said, "like 'What's the risk that an orca will get hit by a rotating blade?'"

Because none of these questions had been asked before, Copping and her team would initiate a study and months later return to NOAA with an answer. Then NOAA would raise another question. "It went on and on," said Copping. "After months of looking at the statistical possibility of an orca getting struck by a rotating turbine, we concluded the risk is very low. First, these tide devices sit about 150 feet deep. Marine mammals spend most of their time in the top 100 feet. Also, whales are smart. They're aware of what's around them and they're unlikely to stick their head intentionally into a moving turbine, even if it's moving slowly, as these devices typically are. Our studies show that if a whale gets struck, it's most likely going to be an adolescent male whose curiosity is getting the best of him. Even then, the likelihood that a collision would be fatal is very, very small."

≈

"I'm afraid this is all going to end in tears," said Canadian physical oceanographer Chris Garrett, who served on a recent U.S. Department of Energy Commission to assess marine energy potential. The commission's report concluded that the United States could get up to one-quarter of its energy from tides. But Garrett doesn't think it's worth it. "There's plenty of energy in the ocean. That's not the problem," he said during one of my several visits to his home near Oak Bay, on Vancouver Island. "The problem is that when you pencil out the costs to extract the small percentage of available energy, and then tally the energy and money that's already been spent studying the subject, it doesn't add up."

Between presentations at the Punta Arenas conference, I mention Garrett's comment to Davies. He takes a few minutes to consider his response. "Garrett is a highly respected scientist," he begins, "but he's looking at only one dimension of the picture. Can you think of a single technology that hasn't been through a phase where the road ahead looks daunting, even impossible? What if we had given up on computers when they were the size of an elephant? Think of how long it took us to develop the airplane. We still don't know to what degree tide energy is going to pay off, but it's far too soon to give up."

Davies's comments reminded me of Magellan, who journeyed with five small ships and 260 men into a vast and unknown ocean-sea. Magellan's greed and cold-heartedness may not be qualities most of us would praise today, but his unflappable courage and persistence in sailing off the edge of the known world are qualities that inspire.

9 Higher Tides

Sea Level Rise from Kuna Yala to Venice

> Not only do the tides advance and retreat in their eternal
> rhythms, but the level of the sea itself is never at rest. . . Today a
> little more land may belong to the sea, tomorrow a little less.
>
> —*Rachel Carson, The Edge of the Sea*

At midday, woodsmoke seeps from the thatched roof of a community hut on Carti Sugtupu, a small Caribbean island. Having canoed from a nearby island and meandered on narrow dirt paths from the beach, I pause outside the hut to wait for Delfino Reyes, a Kuna Yala Indian who has invited me to attend today's Coming-of-Age ceremony. Canoe-loads of Kuna families, the women adorned in piercings and carrying baskets of plantain, coconut, and rice, have paddled from the islands of Sugdub, Tupile, and Ailigandi. These and another 350 low-lying palmed islands make up the San Blas archipelago that speckles Panama's east coast, all belonging to the Kuna Yala.

Two weeks ago, a medicine man began preparing *chicha*, a fermented mead brewed from coffee, coconut, and bananas. With the help of villagers, the ingredients were boiled in cauldrons and poured into large clay pots to mature. The shaman droned a chant as he taste-tested the brew. Only when he announces that the *chicha's* ready can the ceremony be scheduled—in this case for a young girl, Angela, who is a friend of Delfino's family.

Delfino arrives, barefoot, wearing black cotton pants, a pink blouse pleated at the shoulders, and a black felt bowler. He leads me through

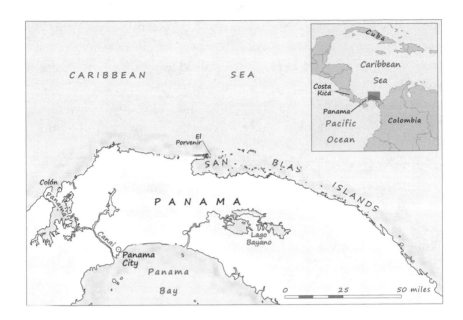

the hut's door into a dirt-floored room about forty feet square, then introduces me to Angela's father and the village chief. The women take Leticia, my interpreter from Seattle, and wrap her in a skirt and *mola*, a reverse-appliqué cotton panel stitched by Kuna women—a craft for which they're known worldwide.

The hut is dim and cool, brimming with people and excitement. Women cluster on one side and men on the other, all nodding and bobbing to the breathy tones of panpipes. At the hut's center, a shaman treats a middle-aged man with smoke from rolled cacao leaves. The shaman passes the burning stub in circles around the man's face, then brings it to his own mouth, takes a long draw, and blows the smoke in staccato beats into the man's nose and ears. The treatment goes on for almost an hour, with no hint of a facial expression from either man, nor notice from the others in the hut.

Angela, the young woman of honor, has spent the last four days in isolation, bathing in seawater, cutting her hair, and dying her body with the genipa plant's black juice, said to identify her with the "new moon that hides its face." Today, arrayed in colorful *molas*, strings of beads tightly wrapped from ankle to knee, and intricate jagua juice tattoos on her face and arms, she's reintroduced to the community, a woman.

Of the hundred or more celebrants, I'm the only gringo. I feel privileged to be here but a little self-conscious. It doesn't help that at six feet tall I stand at least a foot above everyone else. I squat on a low bench, trying to be less conspicuous.

Delfino brings over the chief and half a coconut filled with *chicha*. He chants while dancing in short hopping motions around me. The chief and several others follow in a line, their hands resting on the shoulders of the man in front. After several circles they stop, and Delfino, cupping the coconut shell in his hands, extends both arms and offers me the ceremonial drink, explaining the custom of consuming the whole helping at once. The concoction is black, bitter, and potent. I'm no experienced drinker, but not wanting to offend, I guzzle—and soon stop worrying about being white or tall or anything else. As the day wears on, I doubt anyone else is worrying either.

I had met Delfino two days earlier when I happened upon his one-room museum on this same island, Sugtupu. Leticia and I were the only ones there, but he graciously spent more than an hour discussing Kuna cultural and spiritual history. Speaking mostly in English, he explained that the Kuna migrated north from Colombia and first settled on the San Blas Islands in the early nineteenth century. "We had to fight for our independence," he said, "first with the Dutch, then the French." But their largest struggle, he explained, was with the Panamanian government. In 1925 Panama finally granted the Kuna autonomy and gave them legal rights to the islands and "the mountain," a 120-mile strip of jungle on the east-facing mainland coast. The treaty was codified in 1953, and thereafter the region was called Comarca de Kuna Yala (the region of the Kuna). Unlike the other eleven indigenous Panamanian groups, the Kuna Yala have managed to maintain their autonomy over the years.

One of the many San Blas Islands.

Delfino circled the room, describing the museum's artifacts. There were conch shells, spirit canoes and dolls, clay containers with herbs used for healing, carved wood trinkets for special cures, sketches of the eight spiritual dimensions, and a tablet depicting the *neles*—"seers" or shamans—who have the innate ability to communicate with the spiritual world. Woven baskets hung from the ceiling; *molas* draped the front wall, each with an abstract representation of the world created by Paba and Nana (Father and Mother). "Paba and Nana made everything," Delfino explained. "They watch over us."

Skulls of traditionally hunted animals lined the shelves in descending order of size: the tapir and peccary (wild pigs), the agouti (a rabbit-sized wild rodent), the iguana, and six or seven others, the smallest of which was the squirrel. "Paba brings the tapir down from the mountain about once a year," said Delfino as he picked up the long,

Storm tide flooding in the San Blas Islands, November 2009.

pointed football–sized skull. "They can be six hundred pounds and feed the whole village for a week." In addition to these animals, the Kuna get fish, lobster, and crab from the sea and grow coconuts, plantains, corn, avocados, mangos, red rice, and bananas on "the mountain." What little the Kuna need from outside the islands—gasoline for out-board motors, kerosene for cooking, soap, disposable diapers, and the like—comes once a week by boat from Colón, at the Panama Canal's east entrance, one hundred miles north.

Delfino pulled a cotton hammock from a box and stretched it between his arms. "The hammock is everything to the Kuna people," he said. "We sleep in them, we have babies in them. When someone dies, the body is wrapped in a hammock and hung underground until a spirit takes them to another dimension." Later, I was reminded of Delfino's comments when he took me to visit some island homes. I never saw a chair or couch, only hammocks, often with large brown eyes staring out between the mesh.

As Delfino finished the museum tour, I explained how my research had led me here, having learned that rising tides were a threat to the islands, which hover just waist-high above sea level. Although the tide range in the Caribbean is generally less than a foot, every inch translates into habitable—or uninhabitable—land. Tides several inches higher than normal have already inundated some villages. In fact, if sea level rises two feet in the next fifty years, as conservatively estimated, all but a few San Blas Islands will disappear, forcing thirty-five thousand Kuna to find another home.

The Kuna community has mixed views of the encroaching tide. Some see it as an inevitable fact of modern life. Others, like Delfino, see it strictly as a spiritual matter. "High tides come every year," he said. "I remember them even when I was a child. We protect our houses with sandbags and medicine made by a *nele*. We watch the tide come up the village paths to our doors, carrying visitors from another spiritual dimension. They come to see if everything is in balance. If it is, they go away.

"If it isn't, they stay."

～

The Coming-of-Age ceremony for Angela continues through the night, but I leave after five or six hours, before the sun sets, exhausted and intoxicated. On the ten-minute canoe ride back to my pensione on Tulipe Island, the salty air and spray revive me. Small canoes like the one I'm in—ten to twelve feet long—are crudely hewn from logs felled on "the mountain" and usually skippered by one man with a paddle. Some have outboard motors or sails; all are held together by tar and tin and brads. With rails only a couple of inches above the waterline, these canoes are never dry. Skippers bail with a gourd, constantly.

The canoes are symbolic of how close the Kuna live, literally, to the sea. Like the islands, these small vessels ghost between worlds—not fully afloat nor fully sunken. A wave over the starboard rail can swamp a canoe; a high tide can swamp an island.

A wave is random, but what about the tide? Can we predict the highest tide? How high would it be and how frequently would it occur? If the tides continue to rise, what will the Kuna and other vulnerable populations do?

There's no simple answer to these questions, but we can get a clearer picture by looking at the interactions of three global forces: tide, weather, and long-term sea level rise.

Technically, the tide is an astronomically influenced change in sea level—a predictable event based on known periodic motions of the earth, sun, and moon. The year's tide tables are calculated using these known factors, yet any number of things can trump the predicted tide, both in time and height, including barometric pressure, wind, waves, rain, or storm surge. These events—often called *meteorological tides*—occur on timescales of days, weeks, or months. Over much longer timescales of hundreds, thousands, even millions of years, the world's oceans heave up and down like the tide itself, in sync with the planet's warming and cooling cycles.

I had a firsthand experience of a meteorological tide on the day after Christmas in 2014. Where I live, the tide was predicted to peak at 10.5 feet, the year's highest. I knew a tide like that would flood the road below my home, so I got up early to see it. I walked the beach and waited, but the water seemed to stall a couple of feet below the road. I went home, puzzled.

The next day the *Seattle Times* announced that the region had seen record-high barometric pressure. The correlation with the tide wasn't mentioned, but the high pressure exerted enough weight on the ocean's surface to keep it from reaching its predicted high.

The opposite had occurred a couple of weeks earlier. It was a neap tide, with a predicted high water of 8.5 feet. Onshore winds and low pressure added another two feet, bringing the total to 10.5 feet. *That* tide flooded the road.

A tide-prediction machine used by the U.S. Coast and Geodetic Survey from 1912 to 1965.

William Thomson's 1872 tide-predicting machine is a good visual image of how multiple elements can work together—or not—to create the tide we witness at a beach. Until his time, tide prediction was accomplished mostly by individuals who kept their dubious methods secret. Thomson's invention was the first to integrate a global tide theory into a single tide-predicting machine. His design had ten brass gears, all of different sizes turning at different speeds. Each gear represented a harmonic constituent, in the same way that each kick on Charlie O'Reilly's table represented a pulse from the heavens. As the gears turned, the motion was translated through pulleys and cogs to the machine's far end where it produced a wavy pencil-line on a rotating spool of paper. This, unfurled, became the tide chart.

In the fifty years following Thomson's invention, at least ten similar machines were built, each adding gears and pulleys for further refinement. The first U.S. machine had nineteen gears, India's had twenty-four, and Britain's second machine had forty. The last and largest was a sixty-two-gear machine built for the Germans in 1938. These ten machines were responsible for cranking out the world's tide tables until computers took over in the 1960s.

I haven't seen one of Thomson's machines, but I can imagine its fine-toothed brass gears, tuned as precisely as a clock. These machines were the size of a small car—and as heavy. The earliest models typically sat silently in the back room of some government office. When it was time for an updated tide prediction, someone with strong arms was brought in to turn the crank. The gears came to life, whirring and ticking at varying speeds, each "weighted" according to its tide contribution. The gears representing the solar and lunar days were small, turned fast, and were heavily weighted; the gears for new and full moon were larger and turned more slowly, as did the gears for lunar perigee and apogee. Other gears represented the moon's twenty-seven-day cycle as it travels north and south of the equator, the spring-fall cycle of solar equinox, and the earth's yearly cycle as it moves closer and farther from the sun (perihelion/aphelion).

The most extreme tides, both high highs and low lows, occurred when two or more of these "gears" arrived at the same place at the same time (the equivalent on O'Reilly's table of two or more simultaneous kicks). This happens, for example, every fourteen months when new or full moon (syzygy) coincides with the moon's closest approach to the earth (perigee). Because these two gears are heavily weighted, the event registers as a spike on the rotating spool of paper. As we've learned, this is called a *perigean syzygy* and raises *perigean spring tides*.

The slowest-moving gear on most of these machines represented the 18.6-year cycle of the moon reaching its highest declination (28.5 degrees) north and south of the equator. This is usually the longest cycle used for prediction, as most tides repeat themselves at this interval. The tide chart for 2016, for example, is almost identical to the one for 1998.

If we built a tide-predicting machine that represented all four hundred astronomical tide constituents, it would be huge. We'd need a sizable space for it, such as a large theater stage. While we study the glistening contraption from a soft audience seat, we might notice that even when multiple gears line up, they don't often line up *exactly*. If several gears arrived near the same spot within a forty-eight-hour period, that might be *close enough* to cause a spike on the tide chart. But it's not as close—nor is the spike as large—as it could be.

Fergus Wood, an oceanographer and geophysicist who worked for the U.S. National Ocean Service in the mid-twentieth century, noticed that there were perfect and not-so-perfect alignments, too. In his 1978 book, a ten pound, 550-page tome called *Tidal Dynamics*, he analyzed the highest astronomical tides (HAT) in the 400-year period between 1600 and 1999. The first study of its kind, Wood's analysis revealed that the highest tides varied as much as 40 percent, depending on which "gears" aligned and how closely. In other words, a perigee-syzygy separation of five days might raise a twenty-foot tide, while a separation of five minutes might raise a twenty-eight-foot tide.

Until Wood's time, the highest tides were lumped under the term "perigean springs." Seeing the need for further distinction, Wood introduced several new terms and five new tide categories. At one end of the spectrum, the tide raised by a loose perigee-syzygy alignment of 3–5 days was called an *ordinary spring tide*.

If a perigee-syzygy separation was thirty-six to eighty-four hours, it produced a *pseudo-perigean spring tide*. If the alignment was within ten hours, it resulted in a *proxigean spring tide* (from Latin *proximus*, meaning "nearest"). These happen about every eighteen months.

Wood went on: a perigee-syzygy separation of only five hours raised an *extreme proxigean spring*, something we'd see every ten years. If separation was less than five hours, Wood called it a *maximum proxigean spring*. These happen about every twenty-eight years.

The largest astronomical tide occurs when the separation between perigee-syzygy is less than an hour and coincides exactly or almost exactly with at least one of five other heavily weighted "gears" on our machine: summer or winter solstice, spring or fall equinox, the moon and sun at the same declination, the earth at its closest approach to the sun (perihelion, January 4–6), or the moon at zenith where it's as much as 4,000 miles closer to a point on earth directly below.

What all this means is that the most extreme tides happen when the sun and moon are nearest the earth and as closely aligned as theoretically possible. Wood christened the rarest of these coincidences an *ultimate maximum proxigean spring*. It happened in 1340 CE and will happen again in 3181 CE.

Meanwhile, every year has a highest tide, as does every decade, every century. These vary in height from port to port depending on local resonances, proximity to amphidromic systems, or other shallow-water effects. San Francisco will see an exceptional tide in December 2026 and 2057, New York in November 2038 and December 2052, and Galveston in December 2039 and 2057.

In December 2039, the Kuna Yala will see a tide that's 20 percent higher than today's average spring tide. That tide will inundate many of the islands.

≈

From our theater seat, we might get over our infatuation with the glistening tide machine when we realize that something important is missing. What about all the certain but less-predictable weather influences? To include them, we'd have to add a lot more gears and build in a provision for randomness: today the weather may push the tide earlier and higher, tomorrow it may hold it back, next week it may have no effect at all.

Fergus Wood dedicated much of his book to studying coastal flooding in North America due to the coincidence of large storms and high tides. In the 400-year period of his analysis, he found a hundred such events—thirty of them involving hurricanes—all causing massive flooding and loss of life.

August 1635 was the earliest recorded. "Such a mighty storme of wind and raine as none living in these parts, either English or Indiens, ever saw," wrote Plymouth Governor William Bradford, who had sailed over from Britain on the *Mayflower* fifteen years earlier. "It caused the sea to swell," he continued, "above 20 foote, right up and downe, and made many of the Indeans to clime into trees for their saftie. . . . The signs and marks of it will remaine this 100 years in these parts wher it was sorest."

Another large storm hit the east coast on March 7, 1723, during a proxigean spring tide. The *Boston News-Letter* reported, "The water flowed over the wharf's and into our Streets to a very surprizing height. At Hampton, New Hampshire, the storm caused the great waves . . . to break over its natural banks, and the ocean continued to pour its water over them for several hours."

Wood collected hundreds of newspaper articles describing the severity of these storm-tide events. The headlines themselves are revealing: "Scenes of Destruction at Old Coney Island," announced the *New York Times* on November 25, 1901. "Seas Lashed by Gale—Batter Coast Towns," reported the *Los Angeles Times* on December 18, 1914. "Unusually High Tide Drives Water to Station Entrances in Jersey City," proclaimed the *New York Times* on April 13, 1918. "The water," the article continued, ". . . flooded streets, undermined houses, interfered with ferry traffic, and caused discomfort to thousands of persons. In New Jersey the water came up so high it flooded the waiting rooms of the railroad stations." Other headlines announced: "Pacific County Is Hit by Tide" (Seattle, 1923), "Coast Area Pounded by Rains, Tides" (Oregon, 1933), and "Lower Manhattan Wetted by Tide as Full Moon Pays Us Close Call" (New York, 1953).

In November 1868 Lieutenant Saxby of the Royal British Navy sent a letter to the London press warning of a large tide and probable flooding on October 5 of the following year, 1869. The phenomenon, he noted, was a rare coincidence of perigee-syzygy at a time when the moon was over the equator and the earth was approaching perihelion. Citing a nebulous relationship between weather and the moon's position on the equator, Saxby predicted that a severe storm would accompany the event. "Nothing more threatening can, I say, occur without a miracle," he wrote.

"With your permission," continued Saxby, "I will during September next remind your readers of this warning. In the meantime there will be time for the repair of unsafe sea walls, and for the circulation of this notice throughout the world."

Saxby's warning was picked up by a Halifax, Nova Scotia, resident who alerted the press a week before the event, predicting that a heavy gale would strike at precisely the same time as the extreme high tide.

Saxby was right about the astronomical alignment and large tide, but neither he nor the Halifax resident could have accurately predicted a storm a year or even a week in advance. Yet it happened.

The morning of Monday, October 4, was foggy and calm, reported a Canadian meteorologist, but as the afternoon advanced, "the wind blew in fitful angry squalls," building to hurricane force by evening. When the storm was at its height, "the tide was much above any preceding mark, was rising rapidly and had an hour and half to come." In St. John, the storm, now called the "Saxby Tide," flooded waterfront buildings, destroyed warehouses, carried away whole barns. Up bay, in Cumberland and Hants Counties, water rose above the second floor of homes, drowned hundreds of cattle and sheep, and mangled miles of railroad. Two fishing schooners were torn from their moorings, swept over the dikes, and deposited three miles inland. In Moncton, water levels topped previous records by 6.5 feet. No tide has reached that level since.

The Saxby Tide, as Wood confirmed, was one among many. Had Wood's study included worldwide events, the list would have grown exponentially. The North Sea and the Bay of Bengal, India, are especially prone to these events. Both basins have large tides and frequent storms, which increases the probability of a simultaneous occurrence. And both basins are bordered by low-lying, densely populated coasts that are highly vulnerable to flooding.

Bangladesh, nestled in the upper Bay of Bengal, has seen more of these storm-plus-tide flooding episodes than anywhere on earth. "About eighty major tropical storms form in the world's oceans each year," write oceanographers David Pugh and Philip Woodworth, "and even though only about one per cent of them impact Bangladesh, they result in over half of the world's deaths from cyclones and their associated surges."

If we extended Wood's study to recent years, we would of course include Hurricane Katrina in 2005, Hurricane Sandy in 2012, and Typhoon Haiyan, which made landfall in the Philippines in 2013. All these had record-breaking tide-plus-storm surges: Katrina, 34 feet; Sandy, 12.5 feet; Haiyan, 17 feet.

Sandy made landfall in New York within a couple of hours of spring tide. Fortunately for the city, the storm hit during the lower of that day's two unequally high tides. Had it hit seven hours earlier, the peak surge would have been a devastating four feet higher. As it was, Sandy flooded fifty-one square miles of New York city, or 17 percent of its land mass. The damage was immense. But it was only a rehearsal.

～

Adding weather gears to our tide machine has almost doubled its size. From our theater seat, the contraption is looking rather gangly, with outstretched appendages supporting gears that stop and start, speed up or slow down, to represent the random but significant meteorological tides. Yet something is still missing. How do we account for the slow rise and fall of global sea level?

Because sea level is the underlying platform on which the daily tides perform, perhaps it's best represented as the stage itself—the floor on which the machine rests. If the stage rises, the tide rises; if the stage falls, the tide falls. Because of this, changes in sea level play a role as important as the moon or the weather.

Global sea level usually changes in small increments over large time periods. As hard as it might be to picture the tide racing around the planet at 450 miles per hour, it's even harder to grasp that the ocean platform itself is constantly heaving—as much as 900 feet up and down over cycles of 100,000 years.

Like the Kuna, many of us are watching the tide lap at our door, wondering how high it will get and how soon. Some of us are just waking up to it, but the phenomenon is not new. The tides, riding on the back of global sea level, have advanced and retreated for as long as the oceans have existed. How we look at this depends on the timescale we use as a reference.

If we go back a couple hundred million years, we'd find the earth's continents bunched together in a large island, Pangaea, circled by an enormous sea. The moon was closer, orbiting faster, kicking up extreme tides. The intertidal zone was hundreds of miles wide, rugged, wave-swept.

Tidal friction was already slowing the earth's rotation and pushing the moon away. Over time, due to seafloor spreading and sinking, Pangaea broke apart, and the continents wandered. New coastlines formed. Oceans were closed in or opened up. Atmospheric warming and cooling drove cycles of glaciation, which alternately locked up and released water, changing sea depths by hundreds of vertical feet.

As ocean basins became smaller or larger, deeper or shallower, colder or warmer, saltier or less salty, their resonant qualities shifted. Some became more tuned to the sun and moon, and others less (recall that the Bay of Fundy came into near-perfect resonance only after sea level rose enough to cover Georges Bank, a "recent" event of about 3,000 years ago).

None of these changes were isolated: when one thing shifted, so did others—*so did the whole system*. As continents drifted apart, for example, the total miles of coastline increased, along with the shallow shelves that rimmed them. As the tides rubbed on these, friction increased, further slowing the earth's day and accelerating the moon's retreat. The earth-moon-sun's shifting relationship affected temperature, weather, and, of course, the tides.

In the early twentieth century, a Serbian mathematician and astronomer, Milutin Milankovitch, identified a 100,000-year cycle that correlates with the ice ages. As the earth wobbles, tilts, and elongates over this period—a period that also influences the tides—it receives more or less radiation (heat) from the sun. Winters become longer and summers shorter. Milankovitch reasoned that every 100,000 years the earth reaches a cold or warm peak (a difference of about nine degrees Fahrenheit). At the earth's coldest peak, glaciers crawl down from the poles and, because they lock up so much water, sea levels drop 300–400 feet. That's enough to expose all the world's continental shelves and leave a dry bridge across the Bering Sea. Because water contracts as it cools (and expands as it warms), sea levels drop even farther during a peak cold cycle.

At the earth's warmest peak, ice slinks back toward the poles, refilling the oceans with meltwater. As this chilly water warms and expands, the oceans swell yet more, reclaiming the land they once gave up.

Ice core samples taken in recent years confirm that glaciation has indeed occurred at 100,000-year intervals, at least over the last 2,000 millennia. This doesn't mean a smooth and predictable curve from cold to warm and back again, landing at precisely the 100,000-year mark. Rather, the *Milankovitch cycle*, as it's now called, identifies a general trend, with smaller cycles of warming and cooling within that period.

The last cold peak of the Milankovitch cycle was about 20,000 years ago. At that time, New York and Seattle were buried in mile-thick ice, as was most of Europe, Russia, and the southern regions of South America. Sea level was about 400 feet lower than today. In the years that followed—what's known as the Holocene epoch—temperatures warmed, glaciers retreated, and sea level rose.

In the 1970s, scientists thought the earth's climate may have reached an interglacial warm peak about 6,000 years ago and had entered a new cooling trend. Some conjectured that we'd see another glaciation period within 10,000 years, when ice would again creep north and south from the poles toward the equator.

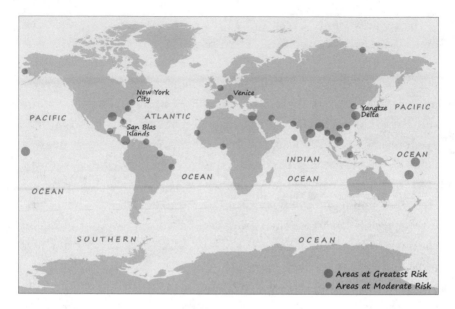

Cities of the world most threatened by sea level rise.

The 1980s, however, brought growing evidence that greenhouse gases, accumulated since the Industrial Revolution, were trapping heat and overriding the earth's natural cycle. Today's climate models confirm this, revealing both conspicuous and nuanced relationships between carbon dioxide buildup, warming air and sea temperatures, larger and more frequent storms, melting ice sheets, and rising sea levels. The interpretations of these models vary, depending on who's financing the research, who's asking the questions, and who's looking at the results. The trend of global climate change, however, is irrefutable.

At stake, among other things, is continued—and even accelerated—sea level rise due to the warming of ocean water (expansion) and the melting of ice sheets and alpine glaciers.

The Intergovernmental Panel on Climate Change (IPCC), published a report in 2013 that predicts a sea level rise of ten to thirty-two inches by this century's end. Because it distills the work of thousands of international scientists, the IPCC report is considered authoritative. Yet many fear its predictions are too conservative. James Hansen, a professor of climatology at Columbia University and outspoken IPCC critic, blames the organization for "scientific reticence," a watering down of predictions by scientists who are overly cautious as individuals, but even more so when working in a group. The IPCC report is "larded with caveats," writes Hansen. The picture coming into focus by satellite imaging, he argues, strongly indicates that ice sheets are already beginning to respond in a nonlinear fashion to global warming. That is, the melting, and consequent sea level rise, will not be steady and predictable but may reach tipping points where a lot of ice melts fast. Hansen claims there's enough information to make it "a near certainty" that continuing our present rate of fossil fuel burning will "lead to disastrous multi-meter sea level rise" in the coming centuries.

"Skepticism is essential to good science," Hansen claims, "but there is a question of degree. We may rue reticence if it locks in future disasters."

⁓

Sea level itself is a slippery concept—not at all easy to distill into definitive measurements or predictions. The sea, in fact, is not *level*. It doesn't lie flat like undisturbed water in a bathtub or pool but piles up in one place and retreats from another. The western Pacific is about three feet higher than the eastern Pacific (due to the northeast trade winds), and the eastern Pacific is ten inches higher than the Atlantic. On the U.S. Atlantic coast, sea levels slope downward from south to north by four or five inches. Different landmasses, due to variations in density and gravitational pull, force seawater this way or that. The oceans are lifted by the Himalayas and Andes and sucked downward by undersea mountain ranges. In the Pacific during El Niño years, when the trade winds relax, seawater that had been pushed westward sloshes back eastward, bringing higher ocean levels to the U.S. west coast.

Seasonal river outflow, due to rain or melting snow, can raise the sea level around river mouths, and the constant cooling and warming of surface water due to upwelling and changes in currents affects sea level too. On the Oregon and Washington coast, the colder water pulled to the surface by spring and summer upwellings cause a twelve- to eighteen-inch drop in sea level.

The land is moving too. Alaska and most of Canada are still rebounding after the removal of weight from the last glaciation, at a rate of about five inches per decade. Other places are sinking, such as Galveston, Bangkok, Tokyo, Jakarta, and Shanghai. In Los Angeles between 1992 and 2000, Huntington Beach and Long Beach rose while Laguna and Hollywood fell.

Measuring sea level, in itself, is tricky. A measurement has meaning only in relation to something else, like a stake in the ground, on a mountain, or a reef that's covered at high tide and exposed at low tide. But these things—reefs and stakes and mountains—are also moving up and down.

Water, as we learned in grade school, may *seek* its own level, but in the big oceans it doesn't *find* it. Hence, when we talk about sea level rise, we're talking about a relative difference between land and sea, each constantly on the move.

⌇

Approaching the San Blas Islands by boat—or canoe, in my case, as I make my way back to my pensione—one has the impression that these islands, many within shouting distance of each other, are so thick with thatched and tin-roofed huts that there's no place to land. Huts are built as close as possible to the water, their beach-side walls soaked at high tide. Outhouses dangle over the water, connected to shore by a couple of loose planks. As much as the Kuna culture is integrated with the sea, it must at times seem cruel to them how the tides rise inexorably, mercilessly, first swelling into the village paths and huts, then retreating, only to swell again.

Several yachts are anchored in the protection of the islands—one from Greece, two from France—probably visiting after several days' passage from the islands of the eastern Caribbean and awaiting their turn to transit the canal to the Pacific. From there, they'll likely embark on the long offshore passages to the Galápagos and South Pacific islands. For now, they're calmly at anchor, enjoying a fresh northeast trade wind that's ubiquitous in all seasons but summer. I'm thankful for the breeze too: it chases away the bugs and stifling heat.

As my canoe gets closer to shore, I hear children, dogs, music, snorting pigs. We land at a small wood dock and clamber ashore. I walk by the one-room cinderblock schoolhouse and a grocery store that sells soap, Pepsi, and canned lunchmeat. Children walk arm and arm, giggling, in underwear and plastic shoes. They call "hello" from inside dim, smoky huts or from between their mother's legs. The island smells of baking bread, tamales, garbage, sewers, kerosene for cooking, and outboard motor exhaust. Colorful clothing—yellows, reds, and blues—luff on lines; bits of paper and plastic blow across courtyards, sticking to bamboo fences. Skinny dogs beg for food. Cats slumber.

The Kuna seem to have a knack for living in many worlds. In a single village, one might find several churches—Mormon, Baptist, and Episcopal—all while the *neles* and village chiefs continue their traditional practices. The *neles* still preside. They lie in hammocks in the community house on most days, chanting and consulting on domestic affairs. A couple might ask permission to be married; someone else asks to travel to Panama City for shopping. If a person is sick, a medicine man may well stand side by side with a Western doctor. One converses with the spirit world and administers cures using wooden dolls and burning herbs while the other prescribes antibiotics and injects penicillin.

The Kuna's apparent ease in accepting what might seem like contradictory views reminds me of a scene in Tracy Kidder's *Mountains Beyond Mountains*. Kidder accompanied physician Paul Farmer as he trekked into Haiti's hinterland, on a mission to improve health. In

Haiti, there's a strong belief in sorcery and the theory and practices of voodoo, especially in the countryside. On one of their treks, they traveled through thick jungle all day, arriving at a small village.

Here, Farmer confronted a woman who a year earlier had accepted his treatment for tuberculosis but on this day was blaming her disease on voodoo. She even knew who was casting the spell on her. "If you believe that," cried Farmer, "why did you take your medicines?" She paused and looked at him as if explaining something to a small child. "*Cheri,*" she finally said, "*eske-w pa ka kon-prann bagay ki pa senp?*" Kidder translated: "Honey, are you incapable of complexity?"

The Kuna appear to have a large capacity for complexity. It's this extraordinary acceptance, this ease of living between two worlds, not struggling with apparent contradictions, that has helped them thrive. In the end, they may be able to exert only a modicum of control over the coming and going of modern cultural tides. They can create community laws, as they have, that restrict foreign ownership of land and business. They have surely demonstrated a successful long-standing resistance to political pressure from countries such as Panama. But they can exert no such control over sea level rise.

The highest tides may indeed bring visitors from another dimension, come to check on the people. As Delfino believes, they leave satisfied that all is well in the community.

No one in the Kuna community would have a problem with the coexistence of this spiritual view of the tide and the modern scientific one that tells them their time on the islands is coming to an end. "We've known this was coming for many years," says Estebancio Castro Díaz, a Kuna lawyer, with whom I spoke by phone. "Of the forty-nine occupied islands, we've already lost three or four. Those communities moved to nearby islands, but soon they'll have to move to the mountain."

Díaz was born and raised on Usdub, one of the larger island communities, and studied business at Victoria Wellington in New Zealand and international law at the office of the United Nations high commissioner for human rights. "We're always thinking about how to maintain our culture's viability while facing economic, health, and political challenges," he says. "Sea level rise is the most important issue we face today, because it means we have to leave our homes and start over. We have to decide which islands should go first and where they should settle, and how. At the present rate, all our islands will be under water in fifty years. It's a different problem than what's facing people living along the coast. They can move upland—to a higher elevation."

There is much to consider in the move. Because the mountain is undeveloped, they will have to plan for infrastructure, density, and potential political challenges with the Panamanian government. "We're looking at house construction too," says Díaz. "Our bamboo huts are not strong, so we're considering other materials and methods of construction so our community structures will last longer."

Not all Kuna believe they should move. While the younger generations seem to be resigned to the facts of sea level rise, the elders, explains Díaz, are more frail, less flexible in their ways, and, understandably, more attached to the islands. The older Kuna, including some of the chiefs who represent the forty-nine communities in the Kuna Congress, are more likely to believe in the tide's spiritual nature. "Some of them think we're being punished," says Díaz, "and that rising sea levels are a sign that we're losing our spiritual equilibrium. They don't want to focus on the move but on finding a better spiritual balance, individually and collectively. There's a lot to think about, and a lot to work out. We're a spiritual people, so every move needs to be considered so that the relationship with our mother and father is not disrupted. Meanwhile, I just hope we're not forced to move before we're ready."

⁓

In the meantime, the Kuna stave off the rising tide by building seawalls of coral, usually harvested by one man in a canoe and stacked at the border of a family home. Even a makeshift wall of four or five inches can make a difference, for now.

Seawall-building has likely been a part of coastal cultures since ancient times. Evidence of it goes back at least four thousand years, to the tidal dockyards at Lothal, India, and the first dikes, fashioned from mud and straw, on China's Qiantang River. In the first century CE, Pliny the Elder reported seeing thousands of earthen mounds strewn across Europe's coastal lowlands. Settlements, called *torps* in Denmark and *warfts* in Germany, were built atop these, safe from the highest tides.

By the tenth century CE, the Dutch, Germans, and Danes had become adept at building embankments and dikes. They learned that by surrounding low marshlands with dikes and pumping the water out with windmills, salty wetlands could be transformed into farmable, livable, dry land. Today, more than half the Netherlands hovers three feet or less above sea level. All that land would be underwater were it not for seawalls.

After a 1953 North Sea storm-plus–spring tide event that breached 1,200 seawalls in Great Britain and broke all previous water level records in the Netherlands, both countries stepped up their defenses. Britain initiated a Storm Tide Warning System and built the Thames Barrier. Completed in 1984, the barrier remains open except when London is threatened by a tide sixteen feet high or higher. When that happens, boat traffic is stopped and four 35,000-ton circular flood-gates swivel into position. The gates, which span the river south of London near the Isle of Dogs, were designed to defend the city against the increasing risks of storms and sea level rise until 2050. The bar-rier closed four times in the 1980s, thirty-five times in the 1990s, and seventy-five times in the first decade of this century.

The Dutch responded to the 1953 storm just as aggressively. Within seventeen days, they established the Delta Committee and charged it with overseeing a new countrywide protection plan. This included the reinforcement of existing seawalls and the construction of a storm tide barrier on the river leading to Rotterdam, Europe's largest port.

lagoon   *sea*

The MOSE tide gates in Venice, Italy.

The barrier, called the Maaslantkering, was completed in 1997 and is expected to be activated once every ten years.

Venice, Italy, which has been challenged by tidal flooding since its founding in the fifth century, began construction in 1995 of its newest protection effort, the Modulo Sperimentale Elettromeccanico (MOSE) gate project. The scheme includes a string of seventy-eight "flaps," each sixty feet long, thirty feet tall, and four feet thick. These lie flat under the three narrow channels that connect the Venetian lagoon to the Adriatic sea. When needed, they rise individually, like the flaps on the wings of an airplane during landing. Once fully raised, they're connected with a spongy gasket (what the engineers call a *Gina gasket*, after the buxom Gina Lollobrigida). With all three channels sealed off, it is hoped that Venice will avert disaster.

~~

To find out more about the MOSE project, I went to Venice in January 2015, a year after my visit to the San Blas Islands, to meet Giovanni Ceccone, acting director of the Consorzio, the organization assembled in 1987 to save Venice from rising tides. In his fifties, Giovanni is energetic, with wavy, swept-back brown hair, wearing glasses, a front-zip sweater, and a scarf. A civil engineer who specializes in oceanography and hydrography, he joined the Consorzio in 1990.

Giovanni flips through a thick document on his desk as he talks, scribbling a signature here and there. When he finishes one, his secretary feeds him another. His conversation, in English, is clear and steady, but he hardly lifts his head for the hour and a half I'm in his office.

Behind Giovanni, through glass walls, I can see the MOSE control room's eighteen-foot-long by twelve-foot-high screen, displaying about eighteen flickering images. Eight specialists sit in front of the screen and monitor its real-time information, such as what each gate is doing; what the weather, tide, and wind are doing; the positions, speeds, and bearings of ships within a hundred-mile radius; the status of power supplies and readiness to activate the gates. There are even indicators that tell the controllers how reliable the screen's information is.

When a large tide is predicted, a cell-phone alert is sent and sirens broadcast throughout the city. There's a low, middle, and high alert. During a low alert, water bubbles from the storm drains in Piazza San Marco, the city's center and lowest elevation. Half the square fills with about eight inches of water, including the entrance to the towering Saint Mark's Basilica. *Passerella*—raised platforms—are set up by public works employees so tourists and locals can walk without getting wet. Cheap plastic boots are sold by street vendors. Otherwise, it's business as usual. These low alerts happen about thirty or forty times a year.

"The MOSE gates will be activated only during the highest alert,"

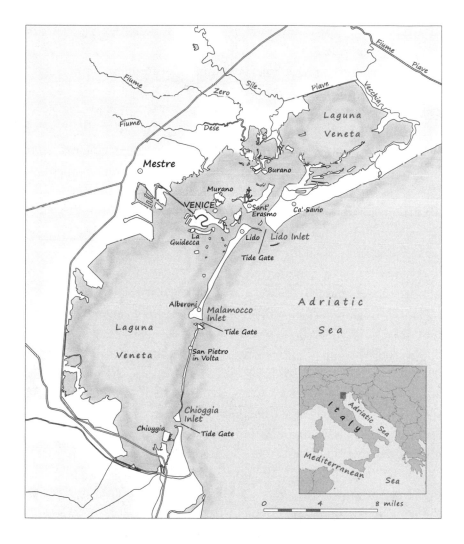

the controller tells me, "which we expect to happen five times a year." Once the signal is given, it takes about thirty minutes to clear boat traffic and raise the gates.

Venice's lagoon, as it's called, is a three-hundred-square-mile estuary protected from the open sea by two low, sandy barrier islands: Lido and Pellistrina. Between these are the three openings: the Chioggia Inlet, the Malamocco, and the Lido. The three-foot Adriatic tides wash in and out of the lagoon through these openings, each less than five hundred yards wide.

"We're hoping the MOSE gates will be fully functioning by 2018," said Giovanni, raising his voice above the boat's engine as we motor out to see the gates under construction. The lead engineer meets us at a locked fence and escorts us down four stories in an elevator to see the underground workings of the project. We walk in a concrete tunnel below the gates that extends all the way across the Lido opening. "We're twenty feet below the bottom of the channel," Giovanni says, his voice tinny against the hard surfaces.

The tunnel is chilly and smells of lime from concrete, and vinegar from the silicone caulking used to embed fasteners and make compartments watertight. Under each "flap" are two four-foot-diameter hinges of gleaming stainless steel, with tidy color-coded cables attached. Giovanni explains that each compartment can be isolated in case of an accident—a flood or a fire.

The gates rest in the down position directly above us. The Lido opening was the widest, so they built an island in the channel's middle and have now completed the gates across one opening, which tested successfully in November 2014. The other gates, as well as two locks designed to allow commercial boat traffic during gate closures, are under construction. Spoils from the dredging are used to construct a marsh inside the lagoon. "Marshes offer a natural resistance to flooding," says Giovanni. "They grow a lot like coral reefs around an atoll. As the land sinks—or the tides rise—the plants and soils build upward, so these marshes become a natural buffer."

The sun is low as we motor back. Our bow slices through oily still-water channels, marked on either side by wood pilings, many of which have been eaten away at the waterline by teredo worms. The basilicas of Santa Maria and Saint Mark's stipple the western horizon. A white egret stands motionless in the gold-lit marsh, hunting for dinner. A grebe, its red eye gleaming, arcs forward and dives.

At a restaurant, Giovanni sips wine and tells the story behind the

MOSE project. After a big tide in November 1966 that covered almost every city street and put Piazza San Marco under three feet of water, the federal government decided to take action. They formed the Consorzio and charged it with saving Venice. "If we don't save Venice, Italy will not be able to stay on the international scene."

From the start, Giovanni's group saw at least three approaches. "We discussed it with the public for ten years, which was slow and tedious, but it resulted in a more integrated solution. We also realized we didn't know everything we needed to know, so we went looking for help. We hired specialists from all over the world."

Venice's tide problem has many dimensions—social, economic, environmental, technical, even philosophical. "We realized early," Giovanni says, "that how you frame the problem makes all the difference in what you discover as the solution. If you're interested in economics, for example, you will model the features of the system that are directly connected with money. If you're interested in beauty, you will construct a model in which you feature beauty. If you're interested in logic, you will phrase the problem in an elegant manner using logic.

"After four or five years of discussion with all sorts of groups, three views surfaced. One was to do nothing—just adapt to the tide, accept the nuisance of water in the city streets and shops. Use pumps, and from time to time raise the banks or demolish a sinking building and rebuild on top, just as we've been doing for five hundred years.

"The second view, similar to the first but more spiritual, is to have faith in nature. This view says, 'If we respect nature, nature will help us. If we allow it, natural processes will shape the solution.' In this view, nature is God."

The third view, explains Giovanni as we're well into dinner, proposed "strong" solutions, involving engineered steel and concrete walls, often called *hard armoring*. Giovanni says these solutions lack resilience. "When a wall like that is breached—and it will breach—then it's catastrophic, because in that system it's either all or nothing."

Giovanni and the Consorzio saw the first option—do nothing—as naïve. "It just brushes the problem off to future generations," he says. "People don't want to change, so they'll tolerate a lot. In our case, the rubber boots keep getting higher; now a lot of us have hipwaders. But if the disturbance is large enough—if water overflows our boots—then it will trigger the search for a new solution. People who wanted to do nothing didn't see that the disturbance was high enough."

In the end, the Venetians decided they could live with occasional flooding in the city's lower elevations, like Piazza San Marco, but they couldn't live with flooding from an extraordinary event like the 1966 tide. They settled on a combination of engineering projects, including the MOSE gates, environmental projects, such as *soft armoring* (beaches designed to absorb tide and wave energy, not block it), and marshes.

"This solution represents, in my mind, a compassionate attitude toward nature because nature is helping us. It's like taking a long trip on a horse. On such a trip, we take care of the horse because the horse takes care of us. We serve it, in a sense, because it serves us. We don't do this because it's a god, but because it helps us reach our destination.

"This is the 'respect nature' approach, which means lots of constraints. You have to travel, for instance, at a pace that is good for the horse, otherwise you will hurt it and you'll end up walking. So it's a process that promotes integrity and compassion."

⩰

The next day I meet Giovanni again, this time to discuss the more technical aspects of the MOSE project. I was surprised to learn that the flood of November 1966, the most extreme in recent history, happened during a neap tide. The tide-plus-storm surge totaled six feet above normal high water. "We were lucky," Giovanni says, "because if the storm had hit during a spring tide, the water would have been three feet higher."

Giovanni figures there's one out of four chances that spring tide and a maximum storm will coincide. To calculate this, he plotted the astronomical tide, hour by hour, over two weeks, then replicated those plots over ten years. He used historical records to add the frequency and likely effects of weather. "Wave energy outside the lagoon can add a foot to the tide," he said, "and if a storm's duration is long, direct rainfall over the lagoon and freshwater discharge from the watershed can increase water level by eighteen inches inside the lagoon." On top of that, he added twenty inches for sea level rise by the end of the century.

Putting this all together, Giovanni figures there's a reasonable chance that Venice will see a nine-foot tide sometime before midcentury. "The MOSE system is designed to accept that possibility," he says.

"We can extend probability models to include worst-case scenarios, but that's unrealistic," he says. "The likelihood of a tide event higher than nine feet in the next hundred years is too low to be considered significant—it's so low that it would be uneconomical to plan for it.

"In the case of exceedance, the tide would overtop the barrier, but it's designed for that, and the exceedance would probably last only about one or two hours without a significant increase in lagoon water level. The system is flexible. It can accept overtopping."

∼

According to my tide chart, high tide in Venice today, January 20, 2015, will be 38.5 inches above normal high tide, one of the year's ten largest astronomical tides. Everyone I see here has a pair of rubber boots; I would have brought mine, but they wouldn't fit in my backpack. Hotels keep spares, however. When I ask, the woman behind the desk produces several pairs to choose from, all sizes and colors. She tells me the water floods the hotel lobby two or three times a year.

Giovanni has arranged for Mattia Balboni, a thirty-three-year-old musician who was born and raised in Venice, to be my guide as the water enters Piazza San Marco. We arrive at 9:00 a.m.; the square is still dry. Pigeons—24,000 of them, says a tourist booklet—waddle on the cobbled square looking for handouts and occasionally taking flight in a rush of wings. Tourists—Japanese, French, British, American—

take pictures of the birds on their shoulders, heads, and outstretched arms. Public works employees unstrap the *passerella* and stack them end to end across the lowest parts of the square. At high tide, these will allow business to continue as usual. Saint Mark's Basilica, which towers over the south end of the square, is the lowest elevation in Venice, thirty-one inches above normal high water, and walking platforms are permanently in place inside the church. Shops and restaurants bordering the square are, at this hour, coming to life.

"I used to come here during high tides in grade school," reminisces Mattia. "Once, I made a steamboat out of a cigar case and watched it sail from one end of the piazza to the other. I grew up with this. I know the barrier project will help control the tide, but if the tide stops, it's going to be very strange for me. I understand it's a problem for some people. Everyone who has a shop has to always watch the tide, and when it comes they have to use a pump and clean the mud afterward. They lose money."

"But for me," Mattia continues, "the tide is part of the city in the same way that the northern lights are part of the Netherlands, or Niagara Falls is part of Canada. When my friends visit from Europe, the tide is one of the things I show them, and they think it's incredible. I'll lose something if we stop the tide, but I know I'll gain something too. I won't have to wake up at two in the morning for fear that high tide is in my living room."

At about 9:30 a.m., just as the shops are opening, the first water bubbles up in front of Saint Mark's Basilica. I stand and watch it surge, recede, then surge again, carrying bits of lettuce, candy wrappers, cigarette butts. The pigeons migrate to the higher elevations at the square's center, and so do visitors, although some, like me, are drawn to the water and play at its edges. A few people remove their shoes, roll up their pants, and wade into it, now about seven inches deep in places. In another hour, the water will pool across the north and east edges of the square.

In China, I saw the tide charge up the Qiantang River as a

Tourists walk on *passerellas* in Piazza San Marco.

twenty-five-foot wall of water; in France, I watched it gallop across the quicksands of Mont Saint-Michel; and in the Arctic, I watched it flood our ice cave more quickly than water fills a bathtub. But seeing it bubble up through a storm drain is a first. If one didn't know it was the tide, it would be a strange and scary sight. It feels like I am looking at the past and future at the same time. Venice has existed for over 1,500 years, but it is only one of many cities built at sea level, and that makes it important to watch.

Mattia and the Venetians, like the Kuna, live with complexity. They love and hate the tide. They know the streets and places in the city with the lowest elevations, where the water will be deepest and hardest to ford. They plan their daily walking routes according to the tide, to avoid water too deep for their boots, just as in other cities commuters plan their driving routes to avoid traffic.

"I remember once when I was about twenty," says Mattia, "I was at a Christmas party and watching the time because I knew the water on the streets near my home would be too deep to pass after 10:00 pm. I left a little late and was walking slowly through the water near my house so it wouldn't overtop my boots. But it was too deep and the water filled my boots. It was so cold that I swore, loudly. I felt bad, especially because it was Christmas night!"

Venetians even advertise places to rent or buy based on the tide exposure. A flat that sits at three feet above normal high water costs more than one that sits at two-and-a-half feet above. "I used to live in a place where the living room and kitchen flooded five or six times a year," says Mattia. "We have sump pumps in our homes, and many of the exterior doors are fitted with bulkheads that can be slid in and caulked for the highest tide."

But mainly, when the tide is high Venetians continue doing what they usually do. Gianfranco Viola, owner of the Aqua Alta bookstore, which sits at twenty-eight inches above normal high water, is accustomed to flooding. "It bubbles up right here and here," he says when I visit, pointing to small holes in the stone floor. "I just lift the crates of books from the floor and set them higher. We stay open. People wade around in rubber boots. I ask them not to splash water on the books or make waves by walking too fast. It's just part of life."

"I have no problem with the water coming in my shop," says a café owner in Piazza San Marco. "We always stay open. The water's dirty, comes from backed-up stormwater systems, so when it ebbs, it leaves a slimy silt—that stinks. We have to hose and squeegee it out afterward. People don't mind standing at my sandwich bar in twelve inches of water, but no one wants to eat a panini while standing in stinky mud."

～

Still in rubber boots, I wander back to my hotel, splashing through puddles left by the ebbing tide. This is Venice's intertidal zone, marked not by seaweed-covered boulders and skittering crabs but by cobbled streets, soaring architecture, and Renaissance paintings. We don't forage for mussels here; we forage for what inspires us about humanity: the brilliance and beauty and creativity for which this city is iconic.

What we harvest in Venice's intertidal zone is hope.

The tide teaches us many things. Giovanni reminded me that even if we stopped all greenhouse emissions today, the momentum of global warming and sea level rise is locked in for at least the next several centuries. That could mean a rise in sea level of fifteen feet—and much more, much faster, if melting ice caps reach a tipping point. No matter how promising the MOSE, London, or Rotterdam barriers are, or the defenses yet to be built by New York, Tokyo, Shanghai, and New Orleans, almost all the world's coastal cities will become intertidal zones in the coming decades.

This shouldn't surprise us, as geological history confirms that the sea's heaving is part of a larger cycle. "Today a little more land may belong to the sea, tomorrow a little less," as Rachel Carson wrote.

But with more than half the world's population living on an ocean coastline, and trillions of dollars invested in vulnerable infrastructure and real estate, we're not all going to buy rubber boots without a fight. The Kuna are lucky because at least they have higher ground to move to. The Maldivians, Bangladeshis, and South Pacific Islanders—all as threatened as the Kuna—don't have that luxury. The Maldivian government, in fact, has already announced its interest in buying a new country.

In the four-hundred-year period covered by Fergus Wood's study, he found so many coincidences of tide and storm that he suspected a cause-and-effect relationship. He even suspected a correlation between tides and earthquakes. His suspicions have not been scientifically proven, but we might allow for the possibility that every tremble, rainstorm, hurricane, or tsunami is connected to the tides and the callings of the moon and sun.

The tide teaches us to live with mystery and complexity. It lives in the body of a mudshrimp, signaling when to swim and when to burrow. It lives in sandpipers, crabs, and whelks. It lives in the spirit of bores, in the prayers of monks. The tide is vibration, music, time.

Acknowledgments

This book would not have made it to press without the guidance and generosity of others, including anthropologists, engineers, ornithologists, fishers, oceanographers of all kinds, translators, sailors, tugboat operators, Coast Guard officers, bookbinders, monks, museum curators, art historians, tide millers, physicists, editors, photographers, illustrators, cartographers, NASA scientists, and surfers.

For science advice, I'm grateful to Emily Carrington, Giovanni Cecconi, Andrea Copping, Graham Daborn, Mike Dadswell, Salludin d'Anglure, Adam Devlin, David DeWolfe, David Drolet, Mike Foreman, David Greenberg, Ivan Haigh, Diana Hamilton, Peter Hicklin, Daniel Kammen, Gita Laidler, Malte Müller, Philip Orton, Julie Paquet, Bruce Parker, Brian Polagye, Richard Strickland, Stefan Talke, Saran Twombly, Sally Warner, Philip Woodworth, Dai Zeheng, and Chen Jiyu.

Oceanographers David Cartwright, Chris Garrett, Walter Munk, Charlie O'Reilly, and Han Zengcui were extraordinarily gracious in enduring my endless (and inexpert) questions. Without them, my dip into the tides wouldn't have gotten past my ankles.

For field research support of all kinds I thank Sergio Andrade, the Avataq Cultural Institute staff, Rupert Baker and the Royal Society staff, Mattia Balboni, Deborah Carr, David Christie, Gareth Davies,

Estebancio Díaz, Shawna Franklin, Joe Gaydos and the SeaDoc Society, Chris Gill, John Goff, Allen Gorden, Vera Grafton, Susy Kist at ORPC, David Klein and the White Construction crew, Greg Long, John Louton, Matthew Lumley and FORCE, Terri McCulloch, Lisa MacKenzie at EMEC, Mary Mijka, Mary Munk, Lukasi Nappaaluk, Pete Naylor, the Orcas Island Library staff, David Plunkett, Amy Porteous, Chris Sauer, Leon Somme, Doug and Laura Tidwell and the Orcas Hotel staff, Sergio Versalovic, Tom Wills, and the monks at Mont Saint-Michel: Jean Gabriel, François-Marie, Laurent-Nicolas, and Sebastian.

Thank you to Heidi Bruce, Lin Gu, Leticia Hopper, Jessica Martin, and Haung Ying for providing translations; special thanks to Leticia, Ying, and Jessica for accompanying me on treks to Panama, China, and France.

I'm immensely grateful to friends and colleagues who read chapter drafts and offered encouragement and advice: Ted Braun, Tom Coleman, Ingrid Emerick, Martha Farish, Pam Loew, Richard Nelson, Barbara Rumsey, Susan and Paul Strong, Lisa West, and Steve White; also, the faculty and my fellow students at Vermont College of Fine Arts, especially Connie May Fowler, Philip Graham, Patrick Ross, and Larry Sutin. My dearest family tolerated with goodwill and humor five years of tide-centric dinner conversations.

The art team's keen eyes and talent brought the book's look and feel together: John Culp, maps and figures; Nicholas Judson, illustrations; Margaret Lanzoni, photo editing; and David Peattie, book design.

Without my editors, I would have bled out long ago: Kris Fulsas, fast and exacting, was onboard from the very beginning; Michael Strong understood *everything* I was attempting to do and, where he found weaknesses, gently but firmly insisted that I do better. Philip Hanrahan reviewed the manuscript when it was first assembled as a whole; Daniel Simon copyedited the final manuscript.

I owe huge gratitude to Laura Blake Peterson, my sharp, talented, and no-nonsense agent at Curtis Brown, for guiding me through the oblique publishing process and introducing me to Barbara Ras at Trinity University Press.

Barbara and her staff—Steffanie Mortis, Sarah Nawrocki, Tom Payton, Lee Ann Sparks, and Burgin Streetman—brought exquisite attention to the myriad publishing details.

Writing a book is not just about writing. It's an intense emotional and intellectual journey. I could not have made landfall on the other side of this book-writing ocean without the loving and intelligent support of my wife, Donna Laslo, and son, Matthew Laslo-White.

Image Credits

All color insert photography © Michael Marten. www.michaelmarten.com
All maps drawn by John Culp.
Page 20, 29, 182, 239, and 242 illustrations by Nicholas Judson.
Page 34, 35, 36, 59, 68, 116, 117, 122, 136, 137, and 228 figures by John Culp.

Page 4. Sheila Kelly
Page 15. © Yva Momatiuk and John Eastcott / Minden Pictures
Page 18. © Laszlo Podor / Getty Images
Page 28. © Parks Canada, Painting: Lewis Parker
Page 48. © Vincent M.
Page 52–53. © Pekka Parviainen / Science Source
Page 54. Wellcome Images / Science Source
Page 80. © Zhu Feng
Page 87. Courtesy of United Kingdom Hydrographic Office
Page 93. Wang Chaoying, 2011 / Imaginechina / AP Images
Page 102. Royal Astronomical Society / Science Source
Page 111. Royal Astronomical Society / Science Source
Page 124. Courtesy the Royal Society
Page 129. © ZUMA Press, Inc. / Alamy
Page 178. © Chris Cheadle / All Canada Photos
Page 187. © Sara Argue / saraargue.com
Page 198. Chris Gill, WestBoundary Photography
Page 240. Composite image: Siemens Energy (*top*); Minesto (*middle left*); Lunar Energy (*middle right*); Tidal Generation, LTD (*bottom*)
Page 255. © jarnogz courtesy of iStock photos
Page 256. © Bear Guerra.
Page 259. National Oceanic and Atmospheric Administration / Department of Commerce
Page 283. Andrea Pattaro / AFP / Getty Images

Glossary

Age of the tide: The time difference between new or full moon and the arrival of maximum spring tides. In most places, the tide's age is 1.5–2 days, but it can be as extreme as plus or minus 7 days.

Amphidromic system: A system around which the tide circles (counterclockwise in the northern hemisphere and clockwise in the southern hemisphere).

Aphelion: A point when the earth, in its elliptical orbit, is farthest from the sun. Currently in early July.

Apogean tide: A smaller tide when the moon is at apogee.

Apogee: The point farthest from the earth in the moon's elliptical orbit.

Barometric pressure: The weight of the atmosphere. Normal pressure at sea level is equal to the weight of a 29.92-inch column of mercury.

Bore: A tide traveling up a river in the form of a wave. Other local names include *eagre, pororoca,* and *mascaret.*

Centrifugal force: The apparent force away from the axis of rotation, exerted on an object moving in a circular or curved path.

Centripetal: The apparent force toward the center of rotation, exerted on an object moving in a circular or curved path.

Continental shelf: The offshore extension of the continent, before the vertical drop into the deep ocean.

Coriolis force: The apparent force exerted on objects moving over the surface of a rotating sphere. On earth, this force causes moving objects to veer to the right in the northern hemisphere and to the left in the southern hemisphere.

Declination: The angular distance of the sun's or moon's orbit north or south of the celestial equator. The sun's declination varies from 23.5 degrees north in summer to 23.5 degrees south in winter. The moon moves through a declinational cycle (north and south of the equator) every 27.21 days, but its maximum declination varies every 18.6 years between 18.5 and 28.5 degrees.

Diurnal tide: Once daily.

Double tide: Two high tides or two low tides with only a small dip in elevation between them. Southampton, England, and Cape Cod are examples of places that have double tides.

El Niño (southern oscillation): A phenomenon occurring in the Pacific every three to seven years in which the trade winds weaken, causing changes in climate, currents, and ocean temperature.

Earth tide: Tidal movements in the solid earth.

Ebb current: Tidal current away from shore or down an estuary or river.

Ecliptic: The sun's apparent annual path, as seen from the orbiting earth.

Entrainment: Adaptation to local environmental factors. From the word "train," meaning being "pulled along" by some influence. In the context of the manuscript, we adapt to a new time zone by being influenced by the different day/night cycle.

Equatorial tides: Tides occurring when the moon is over the earth's equator.

Equilibrium tide: The hypothetical tide produced by solar and lunar forces on a continent-free earth covered completely with deep water.

Equinoxes: The two points where the celestial equator intersects the ecliptic. Also, the times in spring and fall when the sun crosses the equator at these points.

Establishment of port: At a particular port, the interval of time between the transit of the moon and the next high water. Also known as *lunitidal interval*.

Fathom: Six feet (originally the length of rope a sailor can stretch between two spread hands).

Flood current: Tidal current toward shore or up an estuary or river.

Frequency: The number of complete wave cycles during a period of time.

Gravitation (Newton's Universal Law of): All particles are attracted to other particles with a force that is proportional to the product of their masses and inversely proportional to their distance apart.

Harmonic analysis: The calculation of the tide as the sum of multiple harmonics or constituents, each with a different period, amplitude, and phase. Developed by Lord Kelvin in the nineteenth century and used for tide prediction today.

High water: The highest water level reached in a tidal cycle.

Highest astronomical tide (HAT): The highest predictable tide due to astronomical factors.

Interglacial: The warm period between periods of peak glaciation. The present interglacial period, which started about twelve thousand years ago, is called the Holocene.

Internal tides: Tidal waves that propagate below the ocean's surface, where density layers meet. These travel more slowly than surface gravity waves, but they can be 1500 feet high.

Inverted barometer effect: When barometric pressure is high, it depresses the ocean's surface, hence lowering it. When barometric pressure is low, the ocean's surface is elevated.

Knot: A unit of speed equal to one nautical mile per hour.

Low water: The lowest water reached in a tidal cycle.

Lowest astronomical tide (LAT): The lowest predictable tide due to astronomical factors.

Lunar day: The time of the earth's rotation in respect to the moon (twenty-four hours and fifty minutes). Also called *tidal day*.

Mean sea level: Mean hourly water-level heights observed over a given period, often nineteen years.

Mixed tide: Tides influenced by both semidiurnal and diurnal components.

Nautical mile: A unit of distance equal to 1.15 statute miles.

Neap tides: Small-range tides that occur at the moon's first and third quarters.

Overtide: A relatively short-frequency tidal wave over the continental shelf, which has a multiple of an oceanic tide frequency. Given its name because it resembles a musical overtone.

Perigean syzygy: The alignment of earth, moon, and sun (syzygy) during lunar perigee (see below).

Perigean tide: A larger tide that occurs when the moon is at perigee.

Perigee: The point closest to the earth in the moon's elliptical orbit.

Perihelion: A point when the earth, in its elliptical orbit, is closest to the sun. Currently in early January.

Period: The time required for one complete wave cycle, such as the time between the passages of successive wave crests.

Relative sea level: Sea level measured relative to a tide gauge.

Resonance: A vibratory response to stimulation. Tidal resonance occurs when the natural period of oscillation of an ocean basin is similar to the astronomical forcing period, typically around the lunar and solar cycles of twelve and twenty-four hours.

Return period: The average time between events, such as a hurricane, tsunami, or an extreme storm surge.

Sea level: The ocean's level after averaging out short-term variations such as wind waves.

Sea level rise: Increases in mean sea level over time.

Semidiurnal inequality: The height difference between the two high waters and the two low waters of a lunar day.

Semidiurnal tide: Twice-daily tides.

Slack water (or slack tide): For a theoretical tide wave, slack water occurs at the flood or ebb's end, before the current reverses. Depending on the location, however, slack water may occur at different times during the tide cycle.

Spring tides: Large-range tides that occur at new and full moon.

Storm surge: A raised area of water, usually directly below and following the path of a storm.

Storm tide: Storm surge plus astronomical tide.

Syzygy: The alignment of the earth, moon, and sun at the time of spring tides.

Tidal constituent: See *harmonic analysis*.

Tidal range: The difference between a successive high and low water.

Tide species: Another name for tide constituent.

Tides: Periodic changes in water level due, strictly speaking, to astronomical forcing. Changes in water level due to weather are often called meteorological tides.

Tropic tides: Tides occurring semimonthly when the effect of the moon's declination is greatest.

Tsunamis: Waves (also called *seismic waves*) generated by seismic activity— sometimes erroneously called "tidal waves."

Vanishing tide: Periods when local tides conspire to create several hours of a relatively stable sea level. This occurs in Los Angeles, for example, when diurnal inequality is extreme.

Wavelength: The distance from one crest to the next of successive waves.

Notes

FOREWORD

Matthiessen read drafts of all but one chapter of *Tides* before his death in spring 2014.

INTRODUCTION

Page 6. *Three different historical accounts suggest that Aristotle's frustration reached such intensity that he ended his life:* Parker, *The Power of the Sea*, 18.

Page 7. *The Maori of New Zealand believed:* Brunner, *Moon*, 27.

Page 7. *In many cultures, there was a perceived "secret harmony" between the tide and human life:* Frazer, *The Golden Bough*, 39.

Page 8. *Leonardo da Vinci was convinced of the latter:* Needham, *Science and Civilization in China*, 494.

Page 9. *The tides rise and fall constantly on all the world's coastlines:* Measurements of the world's coastlines vary from 200,000 miles to 500,000 miles, depending on the scale of measurement used (smaller scales capture more length because they measure more nooks and crannies, whereas the larger scales tend to straighten the coast out). I used a moderate estimate made by the World Resources Institute from data collected by GIS in 2000.

CHAPTER ONE

Page 13. *Every August, just as the Corophium population peaks, 700,000 semi-palmated sandpipers descend:* Percy, "Keystone Corophium," 1–4.

Page 19. *The Bay of Fundy Tourism Partnership describes for the many tourists three kinds of tides:* Terri McCulloch (director, Bay of Fundy Tourism Partnership), personal communication with the author, 2011.

Page 20. *The flock watches for predators:* Pace, "Mysterious Peregrine Falcon," 36.

Page 21. *Some scientists argue that the birds land here not by plan:* Williams et al., "Flyway-Scale Variation," 886–97.

Page 21. *Hicklin's comment reminds me of a passage in Gary Snyder's book:* Snyder, *The Practice of the Wild,* 109.

Page 21. *Yale ecologist G. Evelyn Hutchinson made the point that evolutionary and ecological processes work together:* Personal communication with Gary Snyder. For further information, see Hutchinson, *Ecological Theater.*

Page 26. *The exposed rocks had looked rich with life under the lowering tide:* Steinbeck, *The Log from the Sea of Cortez,* 49.

Page 26. *To make sense of this complex world:* Ricketts and Calvin, *Between Pacific Tides,* part 5, chap. 11.

Page 26. *The intertidal zone is such a gritty, soupy, fecund environment:* Carr, "Importance of Tides," 1–3; Koppel, *Ebb and Flow,* 279.

Page 27. *In that era, the moon was much closer to the earth:* Dorminey, "Without the Moon," 16–28; Koppel, *Ebb and Flow,* 278.

Page 27. *Before the Acadians arrived in the early 1600s:* Percy, "Dykes, Dams and Dynamos," 2.

Page 28. *Salt marshes are among the most biologically rich habitats on earth:* Percy, "Dykes, Dams and Dynamos," 4; Percy, "Whither the Waters?" 3–9.

Page 28. *But the Acadians saw the low-lying fields as valuable agricultural land:* Percy, "Dykes, Dams and Dynamos," 2.

Page 29. *A 1760 provincial report captures the prevailing sentiment:* Percy, *Fundy Issues* 16 (Spring 1999), 6.

Page 30. *As amphipods (only distantly related to shrimp), their chitinous bodies are divided into many segments:* Percy, "Keystone Corophium," 1–2.

Page 30. *Two generations of Corophium are produced each year in the upper bay:* Percy, "Keystone Corophium," 5–6.

Page 31. *They compensate for predation by spawning millions of offspring:* Percy, "Keystone Corophium," 7.

Page 31. *Even single-celled commuter diatoms have internal tide clocks:* Palmer, *Biological Clocks,* 46.

Page 32. *One of the first scientific reports on biological rhythms:* Palmer, *Biological Clocks,* 75.

Page 33. *Biological clocks can lay dormant, too:* Palmer, *Biological Clocks,* 75.

Page 36. *If their eggs are released during the three days of highest spring tides:* Peer et al., "Life History," 370.

Page 36. *The California grunion, a fish the size of a herring, also has an internal pacemaker:* Palmer, *Biological Clocks*, 114.

Page 38. *With La Rance as a model:* Gordon and Longhurst, "Environmental Aspects," 38–45.

CHAPTER TWO

Page 44. *The forty-five-foot tidal range at the Bay of Mont Saint-Michel:* This is the maximum tide range.

Page 48. *If I were standing here sixty thousand years earlier:* Modern man is considered to have emerged out of Africa about two hundred thousand years ago and migrated to Europe about sixty thousand years ago.

Page 49. *In The Moon: Myth and Image, Jules Cashford writes:* Cashford, *Moon: Myth and Image*, 154.

Page 51. *In the ascent to heaven after death:* Brunner, *Moon: A Brief History*, 30.

Page 51. *The earliest known evidence of a moon calendar:* Cashford, *Moon: Myth and Image*, 17.

Page 57. *The oldest hard evidence of awareness and practical use of the tide:* Cartwright, "On the Origins," 107.

Page 57. *Tides are not mentioned in the Bible:* Parker, *Power of the Sea*, 9–12; Boyle, "In the Heart of the Sea," 17–27.

Page 60. *A classic Chinese text of the first century BCE speculates:* Needham, *Science and Civilization in China*, 488.

Page 60. *The tide's importance to Native peoples of North America is evident:* Robinson, *Raven the Trickster*; Armond, "Raven and the Tide Woman," 42. Variations of this myth also occur in Salish, Tsimshian, and Haida myths.

Page 61. *Plato believed the earth was a large animal:* Parker, *Power of the Sea*, 14.

Page 61. *The Roman geographer Pomponius Mela wrote:* Cartwright, "On Origins of Knowledge," 113.

Page 61. *Leonardo da Vinci jotted in his notebook:* Wylie, *Tides and the Pull of the Moon*, 18.

Page 62. *Da Vinci was so taken by this idea that he tried to calculate the size of the world's lung:* Needham, *Science and Civilization in China*, 494.

Page 62. *The tides appear in the sacred texts of the Vedas, the Puranas, and the Upanishads:* Koppel, *Ebb and Flow*, 21.

Page 65. *An obvious relationship existed between lunar cycles and women's reproductive cycles:* Cashford, *Moon: Myth and Image*, 202.

Page 65. *James Frazer writes in The Golden Bough:* 39–40.

Page 66. *Two thousand years later, Dickens wrote in David Copperfield:* 293.

Page 66. *Of the many legendary stories of loss:* Bouet and Desbordes, *Chroniques Latines du Mont Saint-Michel, IXe–XIIe Siècles*, 324–25.

Page 68. *This happens about four times a year:* The full and new moon each occur at perigee about twice a year.

Page 69. *Without the laws of planetary motion:* These cycles are discussed in more detail in later chapters, but see Gamow, *Gravity*, 97–98, for a discussion of the precession of the equinoxes, a cycle of 25,800 years. It should be noted that any wobble of a large celestial body will influence tides on earth.

Page 70. *"For we who inhabit the various coasts of the British Sea know . . .":* Cartwright, David, "On the Origins of Knowledge of the Sea Tides from Antiquity to the Thirteenth Century," p 121.

CHAPTER THREE

Page 77. *There are in the estuaries of many rivers broad flats of mud:* Darwin, *Tides and Kindred Phenomena*, 59.

Page 78. *They were called nong'chao'er, or "tide players.":* Gernet, *Daily Life in China*, 194–97.

Page 81. *This intimate knowledge evolved into the world's first tide table:* Cartwright, "On Origins of Knowledge of Sea Tides," 122–23.

Page 82. *It's been said that a tidal bore is a sonic boom traveling upriver:* Pugh and Woodworth, *Sea-Level Science*, 148.

Page 82. *A fifth-century BCE story describes how a young warrior named Wu Zixu:* The story of Wu Zixu was told to me by Dr. Chen, but it appears in numerous forms in literature. See, for example, Needham, *Science and Civilization in China*, 485.

Page 83. *He is mentioned in a second-century BCE poem by Mei Sheng:* Allee, "Qiantang Tidal Bore," 61–71.

Page 83. *He wrote . . . [Lu Zheshin's enlightenment poem]:* From *Outlaws of the Marsh*, translated by Huang Ying.

Page 84. *"The hymn is significant":* I spoke to Dr. Richard McBride, who now teaches at Brigham Young University, Hawaii, over the phone.

Page 85. *A hydrographer for the British Admiralty, Captain Moore sailed:* For the full text, see Moore, *Report on the Bore of the Tsien-Tang Kiang*.

Page 87. *An observer reported:* "The Bore passed with a loud roar . . . ": For the full text, see Moore, *Report on the Bore of the Tsien-Tang Kiang.*

Page 90. *One of the many challenges then—as now—was how to protect the foot of the dike:* Dr. Han described much of this in conversations with the author. Additional information can be found in Zeheng and Wei, "Fluvial Processes," 5–11.

Page 90. *Where damage occurred, the official and his family were held responsible:* Bingrao, *Tidal Bore.*

Page 92. *The Shanghai Star reported the next day: Shanghai Star,* September 9, 2002.

Page 92. *Just a month before my visit, Liu Taiquan:* See *Chongqing Evening Paper,* August 3, 2007, translated by Ying.

CHAPTER FOUR

Page 97. *The Royal Society of London . . . Sir Isaac Newton's* Principia Mathematica*:* The full title is *Philosophiae Naturalis Principia Mathematica* (*Mathematical Principles of Natural Philosophy*).

Page 103. *Knowing how unpopular his proposal would be, he wrote in the preface of his* Book of Revolutions*:* The full title is *De Revolutionibus Orbium Caelestium* (*On the Revolution of Celestial Spheres*).

Page 104. *"The natural soul of man," he wrote:* Koestler, *Watershed,* 40–41.

Page 104. *Kepler's first book,* The Sacred Mystery of the Cosmos*:* The full title is *Mysterium Cosmographicum: A Forerunner to Cosmographical Treatises, containing the Cosmic Mystery of the admirable proportions between the Heavenly Orbits and the true and proper reasons for their Numbers, Magnitudes, and Periodic Motions, by Johannes Kepler, Mathematicaus of the Illustrious Estates of Styria, Tuebengen, anno 1596.* Koestler, *Watershed,* 44.

Page 105. *Textbooks tell us that Kepler's three laws of planetary motion were pillars in Newton's theory of gravity:* Kepler's three laws of planetary motion are (1) The path of the planets about the sun is elliptical in shape, with the center of the sun being located at one focus (The Law of Ellipses), (2) An imaginary line drawn from the center of the sun to the center of the planet will sweep out equal areas in equal intervals of time (The Law of Equal Areas), and (3) The ratio of the squares of the periods of any two planets is equal to the ratio of the cubes of their average distances from the sun (The Law of Harmonies).

Page 105. *The word "magnet" comes from lithos magnetis:* Gribbon, *Fellowship,* 7.

Page 106. *The British astronomer William Gilbert published De Magnete in 1600:* The full title is *De Magnete Magneticisque Corporibus, et de Magno Magnete Tellure; Physiologia Nova, Plurimis et Argumentis et Experimentis Demonstrata,* usually translated as *On the Lodestone and Magnetic Bodies and on the Great Magnet the Earth.* Gribbin, *Fellowship,* 11.

Page 106. *Kepler's concept of forces, or "soul," was radically new:* Radically new for the Western world, that is. In the eleventh century, there is a Chinese reference to the moon's *ching,* or radiating "seminal virtue." Needham, *Science and Civilization in China,* 492.

Page 108. *During those years, he confided to a friend, "I feel immense sadness":* Sobel, *Galileo's Daughter,* 317, 346.

Page 109. *Of this tragedy, the historian Thomas Kuhn wrote:* Kuhn, *Copernican Revolution,* 199.

Page 110. *"That concept," wrote Galileo in* Dialogue, *"is completely repugnant to my mind":* Koestler, *Watershed,* 170–201.

Page 113. Which is the thing we want: Wylie, *Tides,* p 20.

Page 113. *He [Newton] wrote in a famous letter to Richard Bentley:* Koestler, *Watershed,* 344; Burtt, *Metaphysical Foundations of Modern Science,* 266.

Page 114. *In his "System of the World," Hooke claimed that all bodies have "an attraction or gravitating power":* Gleick, *Isaac Newton,* 118.

Page 115. *We can also envision the moon orbiting the earth about every twenty-nine days:* The lunar month, called a *synodic month,* is actually 29.53059 days, or 29 days, 12 hours, 44 minutes.

Page 117. *The earth is as rigid as a steel ball:* Cartwright, *Tides: A Scientific History,* 139–41.

Page 118. *The tide's daily squeezing and releasing of the earth:* Wylie, *Tides and the Pull of the Moon,* 118–19.

Page 120. *The moon is only 239,000 miles from earth:* This is the average. It's 225,622 miles away at perigee and 252,088 miles away at apogee.

Page 120. *This difference in distance is the reason the sun's tidal influence is only half that of the moon's:* Usually 46 percent.

Page 120. *When the moon is half full, either waxing or waning, it's 90 degrees from the sun:* Due to the sun's influence, the bulge that the moon raises on earth, which one might expect to be directly under the moon, shifts. Sometimes it's ahead of the moon (called *priming*) and sometimes behind (called *lagging*).

Page 121. *In a month's time it darts like a hummingbird:* The distance the moon travels north and south of the equator varies but takes 27.32 days to complete a round-trip journey, called an *anomalistic* month.

Page 123. *No one knows for certain who prepared the mask:* Keynes, *Iconography of Sir Isaac Newton to 1800*, 75–76.

Page 125. *I like what his biographer, John Maynard Keynes, said of him:* Gleick, *Isaac Newton*, 188.

CHAPTER FIVE

Page 129. *I had met Long a couple of years earlier:* Long, "Taming the Silver Dragon," 16–27.

Page 131. *In the fourth century* BCE, *Aristotle noticed how large ocean waves traveled beyond their stormy birthplace:* Parker, *Power of the Sea*, 103.

Page 131. *Leonardo de Vinci, in the fifteenth century, studied waves:* Parker, *Power of the Sea*, 103.

Page 132. *Newton hinted that the tide had wavelike qualities:* Cartwright, *Tides*, 41.

Page 132. *In 1678 Francis Davenport, an Englishman who was surveying the Red River:* Koppel, *Ebb and Flow*, 125.

Page 132. *Halley called it "wonderful and surprising":* Koppel, *Ebb and Flow*, 125.

Page 132. *Newton didn't call the Tonkin tide a wave:* Cartwright, *Tides*, 41.

Page 134. *The general rule was unraveling at home, too:* Pugh and Woodworth, *Sea-level Science*, 6.

Page 134. *These humps—what we experience as high tide:* The earth is actually turning under these bulges, so as we stand on the beach the tide is not coming toward us, but we are going toward it.

Page 134. *To further complicate matters, they do all this on an orb spinning:* It rotates at this speed at the equator. North or south of the equator, the speed is less.

Page 135. *In 1738 the Paris Académie, still unsatisfied with the progress of tide theory:* Parker, *Power of the Sea*, 27; Cartwright, *Tides*, 59.

Page 135. *The prize was shared two years later by scientists from four countries:* Contest winners were Antoine Cavalleri from France, Leonhard Euler from Switzerland, Colin Maclaurin from Scotland, and Daniel Bernoulli from Holland. Parker, *Power of the Sea*, 27.

Page 138. *Among other things, these initial observations revealed:* Cartwright, *Tides*, 121.

Page 141. *Benjamin Franklin, who was fascinated by the ocean, wrote:* Parker, *Power of the Sea*, 106.

Page 141. *Tsunamis, storm surges, and tides are waves, too:* Bascom, *Waves and Beaches*, 40–91.

Page 142. *Katrina's storm surge, which hit the Gulf Coast in 2005:* Masters, "US Storm Surge Records."

Page 142. *Sandy's [storm surge], which made landfall in New York and New Jersey in 2012:* I don't mention the period of a storm surge because it's usually a single wave, and *period* wouldn't apply.

Page 142. *Some waves feel the bottom and some don't:* Wave behavior is complicated. According to University of Washington oceanographer Richard Strickland, "In 'deep' water, the speed of the wave is determined only by the period or wavelength of the wave, not by the depth. At a depth of half the wavelength (L/2), friction begins to act on the wave and it begins to slow down. Its speed is governed by a complex combination of the effects of both period/wavelength and depth. Waves propagating in water between L/2 and L/20 are called *transitional* waves. 'Very shallow-water waves' are defined as those propagating in a water depth less than one-twentieth the wavelength (L/20). This is the depth at which the speed of the wave is determined only by the depth of the water, not by the period or wavelength."

Page 145. *Traditional Polynesians made an art out of remembering waves:* Genz, "Wave Navigation in the Marshall Islands," 236–40.

Page 146. *"First thing I hear," Melekian says:* Melekian, "Road Agent," 19–20.

Page 147. *He's perhaps best known for his writings on the philosophy of science:* Whewell wrote many books, but the two he is best known for are *History of the Inductive Sciences* (1837) and *The Philosophy of the Inductive Sciences, Founded Upon Their History* (1846).

Page 147. *He wrote sermons, lectured on Gothic architecture:* Snyder, *Philosophical Breakfast Club,* 251, 255.

Page 147. *In 1833, after Samuel Taylor Coleridge complained that "natural philosopher" was the wrong term:* Snyder, *Philosophical Breakfast Club,* 1–3.

Page 147. *If "philosopher" was "too wide and lofty a term," Whewell said:* Science entered the English language in the Middle Ages as a French importation, synonymous with *knowledge.* Ross, "Scientist: The Story of a Word."

Page 147. *Prior to Whewell's involvement with tide studies:* Woodworth, "Three Georges and One Richard Holden," 19.

Page 148. *Bede wrote: "In a given region the moon always maintains whatever bond of union with the sea it once formed":* Cartwright, David, "On the Origins of Knowledge," p 121.

Page 148. *True to his wordsmithing instinct, he called the pursuit tidology:* Reidy, "Flux and Reflux of Science," 170.

Page 148. *While Whewell wrote instructions for observers, Beaufort ensured the cooper-*
ation of more than four hundred Coast Guard stations: Cartwright, *Tides*, 114.

Page 149. *Whewell wrote to a friend:* Deacon, *Scientists and the Sea*, 260.

Page 150. *The Astronomer Royal, George Airy (1801–1892), ridiculed the notion:*
Cartwright, *A Scientific History*, 115.

Page 151. *Captain Fitzroy, master of the HMS Beagle:* Koppel, *Ebb and Flow*, 167.

Page 151. *George Airy agreed:* Koppel, *Ebb and Flow*, 168.

Page 151. *In an 1847 lecture to the Royal Society, he admitted:* Cartwright, *Tides*, 118.

Page 152. *French mathematician Pierre-Simon Laplace:* Laplace became a count
of the First French Empire in 1806 and was named a marquis in 1817, his full
name becoming Pierre Simon, marquis de Laplace.

Page 152. *Using calculus and trigonometry, he [Laplace] developed several highly*
sophisticated equations: Cartwright, *Tides*, 70–71. These are the three fundamental equations (6, 7, and 9) for the dynamics of fluid motion on the surface
of the rotating earth.

Page 153. *These circular systems reminded Harris:* Hamilton, *Encyclopedia of*
Ancient History, 377; Golden, *Children and Childhood in Classical Athens*, 23.

CHAPTER SIX

Page 160. *Physicist Richard Feynman said that friction wears out our leather shoes:*
This was at the fifteenth annual meeting of the National Science Teachers
Association, New York City, 1966.

Page 161. *Thus, the moon, which causes the tide, is in turn pushed away by the tide:*
Newton's law of angular momentum states that the total energy of a physical
system is always conserved. In other words, in a physical system, energy is
neither lost nor gained. In the earth/moon rotating system, tidal friction
acts as a brake on the earth's rotation. As the earth's rotation slows (losing
energy), the moon must, according to Newton's law, absorb that energy.
Consequently, it speeds up and spirals into an increasingly distant orbit.

Page 162. *The Coast Salish people called these narrows Skookumchuck, or "fast*
water": Koppel, *Ebb and Flow*, 135.

Page 162. *Continental shelves vary in width:* Rachel Carson, *The Sea Around Us*,
59–61.

Page 166. *In the first century CE, Pliny described this "lagging":* Garrett and Munk,
"Age of the Tide and 'Q' of the Oceans," 494.

Page 166. *Whewell . . . called this phenomenon the age of the tide:* Garrett and
Munk, "Age of the Tide and 'Q' of the Oceans," 494.

Page 166. *Worldwide, the tide's age varies:* Garrett and Munk, "The Age of the Tide and the 'Q' of the Oceans," 493.

Page 171. *One of the earliest accounts on this coast:* Koppel, *Ebb and Flow*, 145.

Page 172. *Miraculously, Captain Vancouver must have sailed through:* Koppel, *Ebb and Flow*, 146.

Page 172. *One of the rock's early victims:* Koppel, *Ebb and Flow*, 146.

Page 172. *The rock eventually destroyed or severely damaged more than twenty-five large ships:* Wood, *Tidal Dynamics*, 97–98.

Page 173. *By the twentieth century it was decided that Ripple Rock had to go:* Wood, *Tidal Dynamics*, 97–98.

Page 173. *Today the North Sea Pilot gives the very same warning for the Pentland Firth:* Rachel Carson, *The Sea Around Us*, 117.

Page 179. *The effect of global tidal friction actually acts as a brake on the earth's rotation:* Chan et al., "Late Proterozoic and Paleozoic Tides," 100.

Page 179. *Scientists believe that 400 million years ago:* Chan et al., "Late Proterozoic and Paleozoic Tides," 100.

Page 179. *Richard Brathwaite didn't know that time was slowing down when he wrote:* Richard Brathwaite, *The English Gentleman*, 1630.

Page 180. *But less than a decade after the publication of Principia, Edmund Halley:* Brosche, "Understanding Tidal Friction," 1–6; Koppel, *Ebb and Flow*, 272.

Page 181. *Another century passed before the thread was picked up again:* Gied, "Tidal Friction," 1–4.

Page 181. *In his 1889 book* The Tides and Kindred Phenomena, *he hypothesized:* Cartwright, *Tides*, 261.

Page 181. *The day was only a few hours long:* Darwin, *Tides and Kindred Phenomenon*, 281–82.

Page 181. *Darwin's theory of the moon's origin, known as fission, was disproved:* The newest theory posits that the earth was hit 4.5 billion years ago by a planetary body the size of Mars. The impact threw debris into orbit around the earth, where it aggregated to form the moon.

Page 181. *"Corals," he wrote in 1963, "lay down daily rings":* Wells, "Coral Growth and Geochronometry," *Nature*.

Page 182. *Meanwhile, another source of proof appeared:* Cartwright, *Tides*, 236.

Page 182. *In the 1970s, physicist Stephen Pompea and geophysicist Peter Kahn:* Kahn and Pompea, "Nautiloid Growth and Lunar Dynamics," *Nature*, October 19, 1978.

Page 183. *Geologists have long known about tidal rhythmites:* Chan et al., "Late Proterozoic and Paleozoic Tides," 100.

Page 183. *In the 1990s, very old rhythmites were discovered in Australia:* Chan et al., "Late Proterozoic and Paleozoic Tides," 100.

Page 183. *Some scientists are critical of this kind of research:* Clark, "Daily Growth Lines," 27–41.

CHAPTER SEVEN

Page 194. *At that time the upper bay, with a documented maximum tidal range of fifty-four feet, six inches:* Canadian Hydrographic Service, *Resolving the World's Largest Tides,* 4.

Page 197. *According to an old mariner's formula, the "Rule of Twelfths":* Most NOAA tide tables reference this formula (e.g., the tables for *Central and Western Pacific Ocean and Indian Ocean,* 2006, xii).

Page 199. *The 1950s and '60s brought a quantum leap:* Sturtevant, *Handbook of North American Indians,* 476–80, 499.

Page 199. *Lukasi's generation knew the last village shaman:* Sturtevant, *Handbook of North American Indians,* 503–5.

Page 199. *To the horror and disbelief of the Inuit:* Bernard Saladin D'Anglure, anthropologist, in a personal conversation with the author, fall 2013.

Page 201. *Since the Inuit first arrived in northeastern Canada:* Sturtevant, *Handbook of North American Indians,* 480.

Page 204. *Less obvious was the varying distance of the earth from the sun:* Koppel, *Ebb and Flow,* 198.

Page 204. *To make sense of all these cycles:* Koppel, *Ebb and Flow,* 199–201.

Page 204. *Thomson is best known for his work in thermodynamics:* Koppel, *Ebb and Flow,* 201–3.

Page 206. *Kim Bostwick and a science team from Cornell University:* Bostwick et al., "Resonating Feathers Produce Courtship Song," 41–52.

Page 207. *"Everything at its most microscopic level," writes physicist Brian Greene:* Greene, *The Elegant Universe,* 15 and 206 (italics mine).

Page 208. *The Pacific Ocean has a natural period of oscillation of about twenty-four hours:* Arbic and Garrett, "Coupled Oscillator Model," 564–74.

Page 210. *The cycle that has the biggest periodic kick:* Arbic and Garrett, "Coupled Oscillator Model," 564–74.

Page 212. *The Adriatic resonates with African winds:* Wylie, *Tides and the Pull of the Moon,* 80–82.

Page 212. *Known as "harbor effect," these responses cannot be accounted for in tide charts:* Wylie, *Tides and the Pull of the Moon,* 44.

Page 212. *In shallow water, overtides come into play:* Parker, *The Power of the Sea*, 35.

Page 212. *The table-leg kick delivered by the 2002 earthquake in Indonesia:* Charlie O'Reilly, retired chief tide officer at the Canadian Hydrographic Service, Bedford Institute of Oceanography, Dartmouth, Nova Scotia, in a personal conversation with the author, 2014.

Page 213. *Oceanographers call this "back effect":* Arbic and Garrett, "Coupled Oscillator Model," 564–74.

Page 213. *Lake Michigan, with its natural oscillation period of thirteen hours:* Wylie, *Tides and the Pull of the Moon*, 46.

Page 214. *Around 2,500 years ago, sea level rose enough to flood Georges Bank:* Koppel, *Ebb and Flow*, 279.

Page 214. *In the 1970s oceanographer Chris Garrett studied the effect of a tide-generating dam:* Arbic et al., "Resonance and Influence," 1–5.

Page 216. *Journalist Will Ferguson described an encounter at a New Brunswick party:* Ferguson, "King of Tides."

Page 216. *In the fall of 2002 the Canadian Hydrographic Service issued a letter:* Shown to the author by Allen Gordon, Nunavik Tourism Association.

CHAPTER EIGHT

Page 219. *When Ferdinand Magellan set sail:* Bergreen, *Over the Edge of the World*, 119.

Page 219. *In the sixteenth century. . . the European map of South America ended just south of Buenos Aires:* Bergreen, *Over the Edge of the World*, 14, 15, 73–85.

Page 221. *Antonio Pigafetta, the fleet's chronicler:* Bergreen, *Over the Edge of the World*, 62–63.

Page 221. *The Pacific tide usually crests a few hours ahead of the Atlantic:* The Pacific and Atlantic oceans have different tidal regimes. The Atlantic responds more to the moon and its cycles; the Pacific responds more to the sun (see chapter 7). Hence they "slosh" around at different phases, meeting at Cape Horn and the Straits of Magellan. At the entrance to the straits, average water levels are about twenty inches higher on the Pacific side than on the Atlantic side.

Page 222. *When Charles Darwin sailed through here on the* Beagle *in 1836:* Moss, *Patagonia*, p 105.

Page 223. *I was there visiting the European Marine Energy Center:* Lisa MacKenzie, EMEC, interviewed by the author, 2014. You can find more information on this organization at www.emec.org.uk.

Page 224. *For comparison, the European Union aims for 20 percent by 2020:*

Worldwide portfolio standards are readily available on the Internet. See the website of the International Energy Agency, www.iea.org.

Page 232. *When Queen Eleanor visited Nottingham:* Freese, *Coal,* 15–42.

Page 232. *In 1700, addressing the insidious smoke that plagued London, essayist Timothy Nourse wrote:* Freese, *Coal,* 35.

Page 235. *Perhaps the earliest modern-day effort:* Clancy, *The Tides,* 147–57.

CHAPTER NINE

Page 253. *Angela, the young woman of honor, has spent the last four days in isolation:* Salvador, *The Art of Being Kuna,* 281.

Page 254. *Unlike the other eleven indigenous Panamanian groups:* Guidi, "Will a UN Climate Change Solution Help Kuna Yala?" 17.

Page 257. *As Delfino finished the museum tour, I explained how my research had led me there:* Ventocilla, *Plants and Animals in the Life of the Kuna,* 21.

Page 257. *Although the tide range in the Caribbean is generally less than a foot:* The San Blas tide averages thirteen inches. Ventocilla, *Plants and Animals,* 21.

Page 258. *The correlation with the tide wasn't mentioned:* Cliff Mass weather blog, "High Pressure Means Low Sea Level," Thursday, Jan. 1, 2015, and "Seattle sets all-time record high atmospheric pressure today," Dec. 30, 2014.

Page 259. *His design had ten brass gears:* Parker, *The Power of the Sea,* 37.

Page 260. *The slowest-moving gear on most of these machines:* As described in earlier chapters, the moon migrates north and south of the equator every 27.33 days, but its peak declination varies with each of these migrations, reaching 28.5 degrees north and south of the equator every 18.6 years.

Page 261. *Fergus Wood, an oceanographer and geophysicist:* Wylie, *Tides and the Pull of the Moon,* 57.

Page 261. *The first study of its kind, Wood's analysis revealed that the highest and lowest tides varied as much as 40 percent:* Wylie, *Tides,* 60.

Page 261. *Seeing the need for further distinction, Wood introduced several new terms:* Wood, *Tidal Dynamics,* 311–20.

Page 262. *It happened in 1340 CE and will happen again in 3181 CE:* These predictions vary, but Wood's can be found in *Tidal Dynamics,* 313.

Page 262. *Meanwhile, every year has a highest tide, as does every decade, every century:* Statistics obtained through phone and email correspondence with Adam Devlin, PhD, Portland State University. These predictions account for average sea-level rise.

Page 262. *In December 2039 the Kuna Yala will see a tide that's 20 percent higher:*
This is figured for the average spring tide for the period 2015–2040 at
Cristobal, Panama, the nearest primary tide station. This information was
researched and provided by Adam Devlin of Portland State University.

Page 262. *Such a mighty storme of wind and raine:* Wylie, *Tides and the Pull of the
Moon*, 62.

Page 262. *The* Boston News-Letter *reported:* Wylie, *Tides and the Pull of the Moon*,
62–64.

Page 263. *The headlines themselves are revealing:* Wood, *Tidal Dynamics*, 40–56.

Page 263. *In November 1868 Lieutenant Saxby of the Royal British Navy sent a
letter:* Wood, *Tidal Dynamics*, 112–13.

Page 263. *Saxby's warning was picked up by a Halifax, Nova Scotia, resident:* Wood,
Tidal Dynamics, 113.

Page 264. *The morning of Monday, October 4, was foggy and calm, reported a
Canadian meteorologist:* The meteorologist was D. L. Hutchinson. Wood, *Tidal
Dynamics*, 113.

Page 264. *In Moncton, water levels topped previous records:* Koppel, *Ebb and Flow*,
222.

Page 265. *"About eighty major tropical storms form in the world's oceans each year":*
Pugh and Woodworth, *Sea-Level Science*, 157.

Page 265. *Sandy made landfall in New York within a couple hours of spring tide:*
Georgas et al., "The Impact of Tidal Phase," 22.

Page 266. *As hard as it might be to picture . . . as much as 900 feet up and down over
cycles of a 100,000 years:* About 390 feet is the accepted difference, globally,
between the last glaciation period and today, but some locations, like Papua,
New Guinea, have experienced a vertical difference of up to 900 feet. Pugh
and Woodworth, *Sea-Level Science*, 304–5.

Page 266. *If we go back a couple hundred million years:* Koppel, *Ebb and Flow*, 278.
(Pangaea was about 225 million years ago.)

Page 266. *The moon was closer and orbiting faster, kicking up extreme tides:* Koppel,
Ebb and Flow, 278.

Page 267. *In the early twentieth century, a Serbian mathematician and astronomer:*
Milankovitch's 100,000–year cycle is a conglomeration of three cycles: 10,000
years, 20,000 years, and 40,000 years. These peak roughly every 100,000
years. Koppel, *Ebb and Flow*, 280.

Page 267. *Milankovitch reasoned that every 100,000 years:* Englander, *High Tide on
Main Street*, 19.

Page 267. *In the 1970s, scientists thought the earth's climate may have reached an interglacial warm peak:* Koppel, *Ebb and Flow*, 280–81.

Page 268. *The Intergovernmental Panel on Climate Change (IPCC) published a report in 2013:* IPCC, *Climate Change 2013: The Physical Science Basis*.

Page 269. *The IPCC report is "larded with caveats":* Englander, *High Tide on Main Street*, 89–90.

Page 269. *Sea level itself is a slippery concept:* For references in this section on sea level variations, see Pugh and Woodworth, *Sea-Level Science*, 252–300.

Page 269. *The western Pacific is about three feet higher than the eastern Pacific:* Englander, *High Tide on Main Street*, 27.

Page 270. *The land is moving too.* Pugh and Woodworth, *Sea-Level Science*, 300–312.

Page 270. *Other places are sinking:* These cities are sinking not from natural processes but from industrial groundwater extraction. Tokyo, for example, sank fifteen feet in the last century. Pugh and Woodworth, *Sea-Level Science*, 308.

Page 270. *In Los Angeles between 1992 and 2000, Huntington Beach and Long Beach rose:* Pugh and Woodworth, *Sea-Level Science*, 302.

Page 271. *In Haiti, there's a strong belief, especially in the countryside, in sorcery:* Kidder, *Mountains Beyond Mountains*, 27.

Page 272. *"If you believe that," cried Farmer, "why did you take your medicines?":* Kidder, *Mountains Beyond Mountains*, 35.

Page 274. *Settlements, called* torps *in Denmark and* warfts *in Germany:* Parker, *The Power of the Sea*, 62 and 240.

Page 274. *The Dutch responded to the 1953 storm just as aggressively:* Pugh and Woodworth, *Sea-level Science*, 328.

Page 285. *He even suspected a correlation between tides and earthquakes:* Wood, *Tidal Dynamics*, xvi.

Sources

GENERAL

Boon, John. *Secrets of the Tide*. West Sussex, UK: Horwood, 2004.

Brunner, Bernd. *Moon: A Brief History*. New Haven, CT: Yale University Press, 2010.

Cartwright, David. "On the Origins of Knowledge of the Sea Tides from Antiquity to the Thirteenth Century." *Earth Science History* 20 (November 2001).

———. *Tides: A Scientific History*. Cambridge: Cambridge University Press, 1999.

Clancy, Edward. *The Tides: Pulse of the Earth*. New York: Doubleday, 1968.

Darwin, George. *The Tides and Kindred Phenomena in the Solar System*. New York: Houghton, Mifflin, 1898.

Deacon, Margaret. *Scientists and the Sea: 1650–1900*. London: Academic Press, 1971.

Defant, Albert. *Ebb and Flow*. Ann Arbor: University of Michigan Press, 1958.

Frazer, James. *The Golden Bough*. New York: Macmillan, 1951.

Koppel, Tom. *Ebb and Flow*. Toronto, ON: Dundurn Group, 2007.

Marmer, H. A. *The Tide*. New York: D. Appleton, 1926.

McCully, James. *Beyond the Moon*. Singapore: World Scientific, 2006.

Needham, Joseph. *Science and Civilization in China*. Vol. 3. New York: Cambridge University Press, 1959.

Parker, Bruce. *The Power of the Sea*. New York: Palgrave Macmillan, 2010.

Pugh, D. T. *Tides, Surges, and Mean Sea-Level*. New York: John Wiley & Sons, 1987.

Pugh, David, and Philip Woodworth. *Sea-Level Science.* Cambridge: Cambridge University Press, 2014.

Smith, Walton. *The Seas in Motion.* New York: Thomas Crowell, 1973.

Tricker, R. A. R. *Bores, Breakers, Waves, and Wakes.* New York: American Elsevier, 1964.

Wylie, Francis. *Tides and the Pull of the Moon.* Brattleboro, VT: Stephen Greene Press, 1979.

CHAPTER 1. THE PERFECT DANCE

Astro, Richard. *John Steinbeck and Edward Ricketts.* Minneapolis: University of Minnesota Press, 1973.

Binkley, Sue. *Biological Clocks.* Amsterdam: Harwood Academic, 1997.

———. *The Clockwork Sparrow.* Englewood Cliffs, NJ: Simon & Schuster, 1990.

Carr, P. A. "The Importance of Tides and Tidal Heating for Habitability." *Astrobiology*, October 25, 2010.

Comins, Neil. *What If the Moon Didn't Exist?* New York: HarperCollins, 1993.

Cornell Lab of Ornithology. *The Birds of North America Online.* Ithaca, NY: Cornell University, 2015. http://bna.birds.cornell.edu/bna.

Dorminey, Bruce. "Without the Moon, Would There Be Life on Earth?" *Scientific American*, April 2009.

Drolet, David, and Myriam Barbeau. "Diel and Semi-lunar Cycles in the Swimming Activity of the Intertidal, Benthic Amphipod Corophium Volutator in the Upper Bay of Fundy, Canada." *Journal of Crustacean Biology* 29 (2009).

Ellis, David, and Luke Swan. *Teachings of the Tides.* Nanaimo, BC: Theytus Books, 1981.

Foster, Russell, and Leon Kreitzman. *Rhythms of Life.* New Haven, CT: Yale University Press, 2004.

Gordon, Donald, and Alan Longhurst. "The Environmental Aspects of a Tidal Power Project in the Upper Reaches of the Bay of Fundy." *Marine Pollution Bulletin* 10 (1979).

Hicklin, Peter, and John Chardine. "The Morphometrics of Migrant Semipalmated Sandpipers in the Bay of Fundy." *Waterbird Society* 35 (2012).

Holt, Robert. "Bringing the Hutchinsonian Niche into the 21st Century." *Proceedings of the National Academy of Sciences*, November 17, 2009.

Hoshi, Motonori, and Okitsugu Yamashita, eds. *Advances in Invertebrate Reproduction 5.* New York: Elsevier Science, 1990.

Hutchinson, G. Evelyn. *The Ecological Theater and the Evolutionary Play*. New Haven, CT: Yale University Press, 1965.

Koukkari, Willard, and Robert Sothern. *Introducing Biological Rhythms*. New York: Springer Science and Business Media, 2006.

Lathe, Richard. "Fast Tidal Cycling and the Origin of Life." *Icarus,* 2003.

Lopez-Duarte, Paola, and Richard Tankersley. "Circatidal Swimming Behaviors of Fiddler Crab Uca Pugilator Larvae from Different Tidal Regimes." *Marine Ecology Progress Series*, August 7, 2007.

Pace, Jessica. "The Mysterious Peregrine Falcon." *American Museum of Natural History, Learn and Teach*, Fall 2002.

Palmer, John. *Biological Clocks in Marine Organisms*. New York: John Wiley & Sons, 1974.

———. *The Biological Rhythms and Locks of Intertidal Animals*. New York: Oxford University Press, 1995.

———. *The Living Clock*. New York: Oxford University Press, 2002.

———. "Time, Tide, and the Living Clocks of Marine Organisms." *American Scientist* 84 (November–December 1996).

Peer, D. L., L. E. Linkletter, and P. W. Hicklin. "Life History and Reproductive Biology of Corophium Volutator (Crustacea: Amphipoda) and the Influence of Shorebird Predation of Population Structure in Chignecto Bay, Bay of Fundy, Canada." *Wetlands Journal of Sea Research* 20 (1986).

Percy, John. "Dykes, Dams and Dynamos." *Fundy Issues, Bay of Fundy Ecosystem Partnership* 9 (1996).

———. "Fundy's Minas Basin." *Fundy Issues, Bay of Fundy Ecosystem Partnership* 19 (August 2001).

———. "Fundy in Flux." *Fundy Issues, Bay of Fundy Ecosystem Partnership* 15 (2000).

———. "Keystone Corophium." *Fundy Issues, Bay of Fundy Ecosystem Partnership* 13 (August 1999).

———. "Living Lightly on Land and Water." *Fundy Issues, Bay of Fundy Ecosystem Partnership* 24 (2003).

———. "Salt Marsh Saga." *Fundy Issues, Bay of Fundy Ecosystem Partnership* 16 (2000).

———. "Sandpipers and Sediments." *Fundy Issues, Bay of Fundy Ecosystem Partnership* 3 (1996).

———. "Whither the Waters?" *Fundy Issues, Bay of Fundy Ecosystem Partnership* 11 (Spring 1999).

Ricketts, Edward, and Jack Calvin. *Between Pacific Tides*. Stanford, CA: Stanford University Press, 1939.

Seidenberg, David. "Lynn Margulis, Radical Biologist." *Jewish Daily Forward*, December 11, 2011.

Slack, Nancy. *G. Evelyn Hutchinson and the Invention of Modern Ecology*. New Haven, CT: Yale University Press, 2010.

Slobodkin, Lawrence, and Nancy Slack. "George Evelyn Hutchinson." *Endeavor* 23 (1999).

Snyder, Gary. *The Practice of the Wild*. San Francisco: North Point Press, 1990.

Steinbeck, John. *The Log from the Sea of Cortez*. New York: Penguin Books, 1941.

Teal, John, and Mildred Teal. *Life and Death of the Salt Marsh*. Boston: Little, Brown, 1969.

Thurston, Harry. *Tidal Life*. Halifax, NS: Nimbus, 1990.

Verchick, Robert. "Steinbeck's Holism." *Stanford Environmental Law Journal* 22, no. 1 (2003).

White, Jonathan. *Talking on the Water*. San Francisco: Sierra Club Books, 1994.

Williams, Tony, Nils Warnock, John Takekawa, and Mary Bishop. "Flyway-scale Variation in Plasma Triglyceride Levels as an Index of Refueling Rate in Spring-Migrating Western Sandpipers (*Calidris Mauri*)." *The Auk* (American Ornithologists' Union) 124 (July 2007).

Interviews

Debora Carr, Nature of Words Writing Services, 2011.

Emily Carrington, Professor of Biology, University of Washington, Friday Harbor Labs, 2012.

Graham Daborn, Director, Acadia Centre for Estuarine Research, Acadia University, Wolfville, NS, Canada, 2014.

Mike Dadswell, Professor of Biology, Acadia University, Wolfville, NS, Canada, 2012.

Diana Hamilton, Associate Professor, Dept. of Biology, Mount Allison University, Sackville, NB, Canada, 2012 and 2014.

Julie Paquet, Wildlife Biologist, Canadian Wildlife Service, Sackville, NB, Canada, 2014.

Jon Percy, Director, Bay of Fundy Ecosystem Partnership, Wolfville, NS, Canada, 2013.

Saran Twombly, Program Director, Population and Community Ecology, National Science Foundation, 2012.

CHAPTER 2. STAR OF OUR LIFE

Armond, Dale. "Raven and the Tide Woman." *Raven*, 2008.

Bely, Lucien. *Mont Saint-Michel*. Paris: Quest France, 2008.

Bouet, Pierre, and Oliver Desbordes. *Chroniques Latines du Mont Saint-Michel, IX–XII Siècles*. Caen, France: Presses Universitaires de Caen, 2009.

Boyle, Marjorie O'Rourke. "In the Heart of the Sea: Fathoming the Exodus." *Journal of Near Eastern Studies* 63 (January 2004).

Brunner, Bernd. *Moon: A Brief History*. New Haven, CT: Yale University Press, 2010.

Campion, Nicholas. *The Ancient World*. Vol. 1 of *A History of Western Astrology*. London: Continuum International, 2008.

Cashford, Jules. *The Moon: Myth and Image*. New York: Four Walls Eight Windows, 2002.

Dickens, Charles. *David Copperfield*. San Bernardino, CA: A Public Domain Book, 2013.

Dolnick, Edward. *The Clockwork Universe*. New York: HarperCollins, 2011.

Duhem, Pierre. *Le Système du Monde*. Paris: Hermann, 1965.

Frazer, James. *The Golden Bough*. New York: Macmillan, 1951.

Gamow, George. *Gravity*. New York: Anchor Books, 1962.

Laird, E. S. "Albumasar and Medieval Tidal Theory." *Isis* 81 (1990): 684–94.

Maxwell-Stuart, P. G. *Astrology: From Ancient Babylon to the Present*. Gloucestershire, UK: Amberley, 2010.

Needham, Joseph. *Science and Civilization in China*. Vol. 3. New York: Cambridge University Press, 1959.

Panikkar, N. K. "The Concept of Tides in Ancient India." *Indian Journal of History of Science* 6 (1971): 636–50.

Robinson, Gail. *Raven the Trickster*. New York: Atheneum, 1982.

Romer, F. E., ed. *Pomponius Mela's Description of the World*. Ann Arbor: University of Michigan Press, 1998.

Sarton, George. *Appreciation of Ancient and Medieval Science During the Renaissance, 1450–1600*. New York: Barnes, 1955.

———. *Six Wings: Men of Science in the Renaissance*. Bloomington: Indiana University Press, 1957.

Scott, Sir Walter. *Redgauntlet*. Lexington, KY: A Public Domain Book, 2013.

Taylor, Richard. *How to Read a Church*. London: Random House, 2003.

Wedel, Theodore. *Astrology in the Middle Ages*. New York: Dover, 2005.

Wright, Thomas. *The Historical Works of Giraldus Cambrensis*. London: G. Bell and Sons, 2013.

CHAPTER 3. SILVER DRAGON

Allee, Stephen. "The Qiantang Tidal Bore." *Asian Art and Culture* 8 (Spring–Summer 1995): 61–71.

Bingrao, Lin. *Tidal Bore: A Magical Dynamic Phenomenon.* Beijing, China: Higher Education Press, 2010.

Chen, Jiyu, et al. "Geomorphological Development and Sedimentation in Qiantang Estuary and Hangzhou Bay." *Journal of Coastal Research* 6 (1990).

Elisseeff, Serge, ed. "Ritual Exposure in Ancient China." *Harvard Journal of Asiatic Studies* 14 (1951).

Gernet, Jacques. *Daily Life in China.* Stanford, CA: Stanford University Press, 1962.

Johnson, David. "The Wu Tzu-hsu Pien-wen and Its Sources." Parts 1 and 2. *Harvard Journal of Asian Studies* 40 (1980).

Lynch, David. "Tidal Bores." *Scientific American,* October 1982.

Maeda, Robert. "The 'Water' Theme in Chinese Painting." *Artibus Asiae* 4 (1971).

Moore, Robert. *Report on the Bore of the Tsien-Tang Kiang.* London: Admiralty Hydrographic Office, 1888.

Moule, A. C. "The Bore on the Ch'hien Thang River in China." *International Journal of Asian Studies* 22 (1923): 135–88.

Needham, Joseph. *Science and Civilization in China.* Vol. 3. New York: Cambridge University Press, 1959.

Roberts, Moss, trans. *Outlaws of the Marsh.* Beijing: Foreign Languages Press, 1995.

Zeheng, Dai, and Jiang Wei. "Fluvial Processes and Regulation Practice of the Qiantang Estuary." *International Research and Training Center on Erosion and Sedimentation,* circular no. 5 (1989).

Zeheng, Dai, and Zhou Chaosheng. "The Qiantang Bore." *International Journal of Sediment Research* 1 (November 1987).

Zhenghai, Song, et al. *History of Oceanography in Ancient China.* Beijing: China Ocean Press, 1989.

Interviews

Dr. Chen Jiyu, Director, Institute of Estuarine and Coastal Research at East China Normal University, Shanghai, 2013.

Dr. Dai Zehang, Professor and Senior Engineer Honorary Director, Zhejiang Provincial Institute of Estuarine and Coastal Engineering Research, Hangzhou, 1997.

Dr. Han Zengcui, Consultant and Professor, Zhejiang Institute of Hydraulics and
　　Estuary, Hangzhou, 2013.

Interviews, often translated and interpreted by Huang Ying, were conducted
　　with numerous fishermen, political officials, engineers, and scientists.

CHAPTER 4. THE LAST MAGICIAN

Ackroyd, Peter. *Newton.* New York: Nan A. Talese, 2006.

Aczel, Amir. *A Strange Wilderness.* New York: Sterling, 2011.

Atkinson, Dwight. *Scientific Discourse in Sociohistorical Context: The Philosophical
　　Transactions of the Royal Society of London, 1675–1975.* Mahwah, NJ: Lawrence
　　Erlbaum, 1999.

Bergman, Peter. *The Riddle of Gravitation.* New York: Charles Scribner's Sons,
　　1968.

Berlinski, David. *Newton's Gift.* New York: Free Press, 2000.

Bryson, Bill, ed. *Seeing Further: The Story of Science, Discovery, and the Genius of
　　the Royal Society.* London: HarperPress, 2010.

Burtt, E. A. *The Metaphysical Foundations of Modern Science.* New York: Dover,
　　2003.

Clarke, Desmond. *Descartes.* New York: Cambridge University Press, 2006.

Cress, Donald, trans. *René Descartes: Meditations on First Philosophy.*
　　Indianapolis, IN: Hackett, 1993.

Diehl, Edith. *Bookbinding: Its Background and Technique.* New York: Dover, 1980.

Dolnick, Edward. *The Clockwork Universe.* New York: HarperCollins, 2011.

Ellis, Aytoun. *The Penny Universities.* Bristol, UK: Western Printing Services,
　　1956.

Ellis, Markman. *The Coffee-House: A Cultural History.* London: Phoenix, 2004.

Feynman, Richard. *Six Easy Pieces.* New York: Basic Books, 1963.

Finocchiaro, Maurice, ed. *The Essential Galileo.* Indianapolis, IN: Hackett, 2008.

Fontes Da Costa, Palmira. *The Singular and the Making of Knowledge at the
　　Royal Society of London in the Eighteenth Century.* Newcastle upon Tyne, UK:
　　Cambridge Scholars, 2009.

Gamow, George. *Gravity.* New York: Anchor Books, 1962.

Gleick, James. *Isaac Newton.* New York: Vintage Books, 2003.

Greene, Brian. *The Elegant Universe.* New York: Vintage Books, 2000.

Gribbin, John. *The Fellowship.* New York: Overlook Press, 2007.

Harris, Jonathan. "The Grecian Coffee House and Political Debate in London,
　　1688–1714." *London Journal* 25 (2000).

Jamison, Kay Redfield. *Touched by Fire: Manic-Depressive Illness and the Artistic Temperament*. New York: Free Press, 1993.

Keynes, Milo. *The Iconography of Sir Isaac Newton to 1800*. Woodbridge, UK: Boydell P. in association with Trinity College, 2005.

Koestler, Arthur. *The Sleepwalkers*. London: Penguin Books, 1959.

———. *The Watershed*. New York: Anchor Books, 1960.

Kuhn, Thomas. *The Copernican Revolution*. Cambridge, MA: Harvard University Press, 1957.

———. *The Structure of Scientific Revolutions*. Chicago: University of Chicago Press, 1962.

Levy, Joel. *Newton's Notebook*. Philadelphia: Running Press, 2010.

Lieb, Julian. "Isaac Newton: Mercury Poisoning or Manic Depression?" *The Lancet* 322 (1983).

Middleton, Bernard. *A History of English Craft Bookbinding Technique*. London: Oak Knoll Press, 1996.

Nadler, Steven. *The Philosopher, the Priest, and the Painter: A Portrait of Descartes*. Princeton, NJ: Princeton University Press, 2013.

Needham, Joseph. *Science and Civilization in China*. Vol. 3. New York: Cambridge University Press, 1959.

Porter, Roy. *London, a Social History*. Cambridge, MA: Harvard University Press, 1994.

Sacks, Oliver. *An Anthropologist on Mars*. New York: Alfred A. Knopf, 1995.

Sarton, George. *Appreciation of Ancient and Medieval Science During the Renaissance, 1450–1600*. New York: Barnes, 1955.

———. *Six Wings: Men of Science in the Renaissance*. Bloomington: Indiana University Press, 1957.

Sobel, Dava. *Galileo's Daughter*. New York: Penguin Books, 2000.

Solomon, Andrew. "Questions of Genius." *The New Yorker*, August 26, 1996.

Steptoe, Andrew, ed. *Genius and the Mind: Studies of Creativity and Temperament*. Oxford, UK: Oxford University Press, 1998.

Whitehouse, David. *Renaissance Genius: Galileo Galilei and His Legacy to Modern Science*. New York: Sterling, 2009.

CHAPTER 5. BIG WAVES

Bascom, Willard. *Waves and Beaches*. New York: Doubleday, 1980.

Brown, Evelyn, et al. *Waves, Tides and Shallow-water Processes*. London: Open University, 1989.

Cartwright, David. "On the Origins of Knowledge of the Sea Tides from Antiquity to the Thirteenth Century." *Earth Science History* 20 (2001).

Genz, Joseph, et al. "Wave Navigation in the Marshall Islands." *Oceanography* 22, no. 2 (2009): 236–40.

Gillespie, Charles. *Pierre-Simon Laplace: A Life in Exact Science*. Princeton, NJ: Princeton University Press, 1997.

Gladwin, Thomas. *East Is a Big Bird*. Cambridge, MA: Harvard University Press, 1970.

Golden, Mark. *Children and Childhood in Classical Athens*. Baltimore, MD: Johns Hopkins University Press, 1990.

Hamilton, Richard. *Encyclopedia of Ancient History*. New York: John Wiley & Sons, 2012.

Long, Rusty. "Taming the Silver Dragon." *Surfers Journal* 18 (October–November 2009).

Masters, Jeffrey. "US Storm Surge Records." San Francisco: Weather Underground, 2015; Atlanta, GA: Weather Channel, 2015. www.wunderground.com/about/jmasters.asp.

Melekian, Brad. "Road Agent: Greg Long's Life in Pursuit." *Surfer's Journal* 18 (2010).

Pomeroy, Sarah, et al. *A Brief History of Ancient Greece*. Oxford, UK: Oxford University Press, 2004.

Preston, Diana, and Michael Preston. *A Pirate of Exquisite Mind*. New York: Walker, 2004.

Reidy, Michael. "The Flux and Reflux of Science: The Study of the Tides and the Organization of Early Victorian Science." PhD diss., University of Minnesota, 2000.

Ross, Sydney. "Scientist: The Story of a Word." *Annals of Science* 18 (June 1962).

Snyder, Laura. *The Philosophical Breakfast Club*. New York: Broadway Books, 2011.

Woodworth, Philip. "Three Georges and One Richard Holden: The Liverpool Tide Table Makers." *Transactions of the Historical Society of Lancashire and Cheshire* 151 (2002).

CHAPTER 6. FAST WATER

Bills, Bruce, and Richard Ray. "Lunar Orbital Evolution: A Synthesis of Recent Results." *Geophysical Research Letters* 26, no. 19 (October 1999).

Bostock, John, and Henry T. Riley. *The Natural History of Pliny*. London: Henry G. Bohn, 1923.

Brosche, Peter. "Understanding Tidal Friction: A History of Science in a Nutshell." *Science Tribune*, December 1998.

Bunbury, Edward. *A History of Ancient Geography Among the Greeks and Romans from the Earliest Ages till the Fall of the Roman Empire.* London: John Murray, 1883.

Calvino, Italo. *Cosmicomics.* Trans. William Weaver. New York: Harcourt Brace, 1965.

Carson, Rachel. *The Sea Around Us.* New York: Oxford University Press, 1950.

Chan, Marjorie, T. M. Demko, E. P. Kvale, C. P. Sonett, and A. Zakharian. "Late Proterozoic and Paleozoic Tides, Retreat of the Moon, and Rotation of the Earth." *Science*, July 5, 1996.

Clark, George. "Daily Growth Lines in Some Living Pectens (Mollusca: Bivalvia), and Some Applications in a Fossil Relative: Time and Tide Will Tell." *Palaeogeography, Palaeoclimatology, Palaeoecology* 228 (November 2005).

Darwin, George. *The Tides and Kindred Phenomena in the Solar System.* New York: Houghton Mifflin, 1898.

Duhem, Pierre. *Le Système du Monde: Histoire des Doctrines Cosmologiques de Platon à Copernic.* Vol. 3. Paris: A. Hermann, 1913–1959.

El-Sabh, M. I., and T. S. Murty. "Age of the Diurnal Tide in the World Oceans." *Geodesy* 13 (1969).

Feynman, Richard. "What Is Science?" Lecture given at the National Science Teachers Association, New York City, 1966. Reprinted in *Physics Teacher* 7, no. 6 (1969).

Garrett, C. J. R., and W. H. Munk. "The Age of the Tide and the 'Q' of the Oceans." *Deep-sea Research and Oceanographic Abstracts* 18 (1971).

Gied, Brian. "Tidal Friction and the Moon as a Failed Scientific Clock." *The Age of the Earth,* Timothy Heaton seminar, 2005.

Gould, Stephen Jay. *The Panda's Thumb.* New York: W. W. Norton, 1980.

Homer. *The Odyssey.* Trans. E. V. Rieu. Baltimore, MD: Penguin Books, 1946.

Kahn, Peter, and Stephen Pompea. "Nautiloid Growth Rhythms and Dynamical Evolution of the Earth-Moon System." *Nature* 275 (October 19, 1978).

Lambeck, Kurt. *The Earth's Variable Rotation: Geophysical Causes and Consequences.* London: Cambridge University Press, 1980.

MacPherson, Hector. "Kant as an Astronomical Thinker." *NASA Astrophysics Data System,* 2001.

Marsden, B. G., and A. G. W. Cameron, eds. *The Earth-Moon System.* New York: Plenum Press, 1966.

Morrison, L. V., and F. R. Stephenson. "Historical Eclipses and the Earth's Rotation." *Science Progress* 83 (2000).

———. "Historical Values of the Earth's Clock Error Delta Time and the Calculation of Eclipses." *Science History Publications* 35 (2004).

———. "Long-term Fluctuations in the Earth's Rotation: 700 B.C. to A.D. 1990." *Philosophical Transactions of the Royal Society of London* 351 (1995).

Munk, Walter. "The Evolution of Physical Oceanography in the Last Hundred Years." *Oceanography* 15, no. 1 (2002).

———. "Once Again: Tidal Friction." *Royal Astronomical Society* 9 (1968).

Munk, W. H., and G. J. F. Macdonald. *The Rotation of the Earth: A Geophysical Discussion*. London: Cambridge University Press, 1960.

Poe, Edgar Allan. *A Descent into the Maelstrom*. Lexington, KY: Booksurge, 2004.

Saunders, Bruce, and Neil Landman, eds. *Nautilus: The Biology and Paleobiology of a Living Fossil*. London: Plenum Press, 1987.

Stephenson, F. R., and M. A. Houlden. *Atlas of Historical Eclipse Maps: East Asia 1500 BC–AD 1900*. London: Cambridge University Press, 1986.

Stephenson, Richard. *Historical Eclipses and the Earth's Rotation*. London: Cambridge University Press, 1997.

Swanton, John. *Haida Texts and Myths*. Washington, DC: Smithsonian Institution, 1905.

———. *Tlingit Myths and Texts*. Washington, DC: Smithsonian Institution, 1909.

Taylor, E. G. R. *The Haven-Finding Art: A History of Navigation from Odysseus to Captain Cook*. London: Hollis and Carter, 1956.

Thompson, Tim. "Recession of the Moon." *The Talk Origins Archive*, December, 1999.

Thomson, Richard. *Oceanography of the British Columbia Coast*. Ottawa: Department of Fisheries and Ocean, 1981.

Vancouver, George. *A Voyage of Discovery to the North Pacific Ocean, and Round the World*. London: John Stockdale, 1801.

Verne, Jules. *Twenty Thousand Leagues Under the Sea*. London: CRW, 2010.

Webb, D. J. "On the Age of the Semi-diurnal Tide." *Deep-sea Research and Oceanographic Abstracts* 20 (September 1973).

Wells, John. "Coral Growth and Geochronometry." *Nature*, March 9, 1963.

Williams, George. "Geological Constraints on the Precambrian History of the Earth's Rotation and Moon's Orbit." *Reviews of Geophysics and Space Physics* 38 (2000).

———. "Tidal Rhythmites: Geochronometers for the Ancient Earth-Moon System." *Episodes* 12 (September 1989).

Winchester, Simon. "In the Eye of the Whirlpool." *Smithsonian*, August 2001.

Interviews

David Cartwright, Author, *Tides: A Scientific History*, 2012 and 2013.
Chris Garrett, Oceanographer, University of Victoria, British Columbia, Numerous interviews, 2009–2015.
Walter Munk, Author and Oceanographer, Scripps Institute of Oceanography, Several interviews between 2010—2014.

CHAPTER 7. BIG TIDES AND RESONANCE

Arbic, Brian, and Chris Garrett. "A Coupled Oscillator Model of Shelf and Ocean Tides." *Continental Shelf Research* 30 (2010): 564–74.
Arbic, Brian, Pierre St.-Laurent, Graig Sutherland, and Chris Garrett. "On the Resonance and Influence of the Tides in Ungava Bay and Hudson Strait." *Geophysical Research Letters* 34 (2007).
Benade, Arthur. *Fundamentals of Musical Acoustics*. New York: Dover, 1990.
Bessler, Jeromy, and Norbert Opgenoorth. *Elementary Music Theory*. Bonn, Germany: Voggenreiter, 2002.
Bostwick, Kimberly, Damian Elias, Andrew Mason, and Fernando Montealegre-Z. "Resonating Feathers Produce Courtship Song." *Proceedings of the Royal Society*, November 11, 2009.
Canadian Hydrographic Service. *Resolving the World's Largest Tides*, by C. T. O'Reilly. Ottawa, ON: Government of Canada Publications Branch, 1983.
Cheney, Margaret. *Tesla: Man Out of Time*. New York: Touchstone, 1981.
DeWolfe, D. L. "Leaf Basin Tidal Study." Report from Discovery Consultants to Nunavik Tourism Association, July–August 1998.
———. "Leaf Basin Tidal Study." Report from Discovery Consultants to Nunavik Tourism Association, June 2001–August 2002.
———. "Simultaneous Tidal Measurements in Ungava Bay and Bay of Fundy." Report from Discovery Consultants to Nunavik Tourism Association, July–September 2000.
Evan-Iwanowski. *Resonance Oscillations in Mechanical Systems*. New York: Elsevier Scientific, 1976.
Ferguson, Will. "The King of Tides." *Maclean's*, November 18, 2002.
Fletcher, Neville, and Thomas Rossing. *The Physics of Musical Instruments*. New York: Springer-Verlag, 1998.

————. *Principles of Vibration and Sound*. New York: Springer-Verlag, 2004.

Garret, Chris, L. F. Ku, D. A. Greenberg, and F. W. Dobson. "Nodal Modulation of the Lunar Semidiurnal Tide in the Bay of Fundy and Gulf of Maine." *Science* 230 (1985): 69–70.

Garrett, Chris, and Walter Munk. "The Age of the Tide and the 'Q' of the Oceans." *Deep-Sea Research* 18 (1971): 493–503.

Greene, Brian. *The Elegant Universe*. New York: Vintage Books, 2003.

Helmholtz, Hermann. *On the Sensations of Tone*. New York: Dover, 1954.

Hersey, George. *Architecture and Geometry in the Age of the Baroque*. Chicago: University of Chicago Press, 2000.

Leder, M., and M. Orlie. "Fundamental Adriatic Seiche Recorded by Current Meters." *European Geosciences Union* 4 (2004).

Mark, David, Andrew Turk, Niclas Burenhult, and David Stea, eds. *Landscape in Language*. Philadelphia: John Benjamins, 2011.

McGhee, Robert. *The Last Imaginary Place: A Human History of the Arctic World*. Chicago: University of Chicago Press, 2005.

Müller, Malte. *A Large Spectrum of Free Oscillations of the World Ocean Including the Full Ocean Loading and Self-attraction Effects*. Hamburg, Germany: Hamburg University, 2009.

Nappaaluk, Mitiarjuk. *Sanaaq*. Quebec, QC: Bibliothèque Nationale du Québec, 2002.

O'Reilly, C. T. *See* Canadian Hydrographic Service.

Ouspensky, P. D. *The Fourth Way*. New York: Knopf, 1957.

Schureman, Paul. *Manual of Harmonic Analysis and Prediction of Tides*. Memphis, TN: General Books, 2010.

Spatz, Hanns-Christof, Franka Bruchert, and Jochen Pfisterer. "Multiple Resonance Damping, or How Do Trees Escape Dangerously Large Oscillations?" *American Journal of Botany* 94 (2007).

Stuckenberger, Nicole. "The Inuit and Climate Change." *Inuit Studies* 34 (2010).

Sturtevant, William, ed. *Handbook of North American Indians*. Washington, DC: Smithsonian Institution, 1984.

Tagliarino, Barrett. *Music Theory*. Victoria, Australia: Hal Leonard, 2006.

Interviews

Saladin D'Anglure, Anthropologist and Writer, 2011.

David DeWolfe, Retired Chief Tide Officer, Canadian Hydrographic Service, Bedford Institute of Oceanography, Dartmouth, NS, 2013.

Allen Gorden, Director, Nunavik Tourism Association, 2012.

David Greenberg, Tide Modeler, Canadian Hydrographic Service, Bedford
  Institute of Oceanography, Dartmouth, NS, 2011.
Charlie O'Reilly, Retired Chief Tide Officer, Canadian Hydrographic Service,
  Bedford Institute of Oceanography, Dartmouth, NS, Numerous interviews
  between 2010—2014.

CHAPTER 8. TURNING THE TIDE

Beretta, Gian Paolo. "World Energy Consumption and Resources: An Outlook
  for the Rest of the Century." *International Journal of Environmental Technology
  and Management* 7 (2007).
Bergreen, Laurence. *Over the Edge of the World*. New York: Harper, 2003.
Blackwell, Richard. "Nova Scotia Bets on Economic Lift From Rising Tidal
  Technology." *Globe and Mail*, August 27, 2013.
Bridges, Lucas. *Uttermost Part of the World*. New York: Rookery Press, 2007.
Chatwin, Bruce. *In Patagonia*. New York: Penguin, 1977.
Clark, James. *The Chronological History of the Petroleum and Natural Gas
  Industries*. Houston, TX: Clark Book Co., 1963.
DeBlieu, Jan. *Wind: How the Flow of Air Has Shaped Life, Myth, and the Land*. New
  York: Houghton Mifflin, 1998.
Fairley, Peter. "Siemens Boosts Its Stake in Tidal Power." *MIT Technology Review*,
  November 2011.
Ferland, John. "Elements of Success in Tidal Energy Development." *Power
  Engineering*, June 2013.
Freese, Barbara. *Coal: A Human History*. New York: Perseus, 2002.
Garrett, Chris, and Patrick Cummins. "Maximum Power from a Turbine Farm in
  Shallow Water." *Fluid Mechanics* 714 (2013).
Harris, Stephen. "Engineers Head to Swansea for Tidal Energy Lagoon World
  First." *Engineer*, May 2013.
Hassan, Garrad. "Preliminary Site Selection: Chilean Marine Energy Resource."
  Inter-American Development Bank, May 2009.
Hills, Richard. *Power from Wind*. Cambridge: Cambridge University Press, 1994.
Hughes, Donald. *An Environmental History of the World*. New York: Routledge,
  2009.
International Trade Administration. *Chile's Renewable Energy and Energy
  Efficiency Market: Opportunities for US Exporters*. Washington, DC: US
  Department of Commerce, March 2013.

Lothrop, Samuel. *The Indians of Tierra del Fuego*. Ushuaia, Argentina: Zagier and Urruty, 1928.

Marine and Hydrokinetic Energy Technology Assessment Committee. *An Evaluation of the U.S. Department of Energy's Marine and Hydrokinetic Resource Assessments*. Washington, DC: National Academies Press, 2013.

McCarthy, Carolyn, et al. *Chile and Easter Island*. London: Lonely Planet, 2012.

Medeiros, Carmen, and Bjorn Kjerfve. "Tidal Characteristics of the Strait of Magellan." *Continental Shelf Research* 8 (1988).

Molina, Mario, and Durwood Zaelke. "A Climate Success Story to Build On." *New York Times*, September 25, 2012.

Moore, Colin. *Windmills: A New History*. Gloucestershire, UK: History Press, 2010.

Moss, Chris. *Patagonia: A Cultural History*. New York: Oxford University Press, 2008.

Parker, Philip. *Fossil Fuel: Webster's Timeline History*. San Diego, CA: ICON Group International, 2009.

Ruckelshaus, William, Lee Thomas, William Reilly, and Christine Todd Whitman. "A Republican Case for Climate Action." *New York Times*, August 1, 2013.

Smith, Diana. *The Tide Mill at Eling*. Southampton, UK: Ensign, 1989.

Thomson, Oliver. *History of Ancient Geography*. Cambridge: Cambridge University Press, 1948.

University of Edinburgh et al. "Marine Energy Development: Taking Steps for Developing the Chilean Resource." British Embassy in Chile, 2011.

World Bank. *The Little Green Data Book*. Washington, DC: World Bank, 2012.

Wrigley, E. A. *Energy and the English Industrial Revolution*. Cambridge: Cambridge University Press, 2010.

Interviews

Sergio Andrade and Sergio Versalovic, Alakaluf Organization, Chile, 2014.

Carlos Barria, Chilean Ministry of Energy, 2014.

Craig Collar, Snohomish County Public Utility District, 2014.

Andrea Copping, Project Manager, Batelle Seattle Research Center, 2014.

Graham Daborn, Director, Acadia Tide Research Institute, 2014.

Gareth Davies, Director, Aquatera, Scotland, 2013 and 2014.

Simon Geerlofs, Marine Science and Policy Analyst, Batelle Seattle Research Center, 2014.

John Goff and Bud Warren, Tide Mill Institute, 2013.

Kiki Jenkins, School of Marine and Environmental Affairs, University of
    Washington, 2014.
Daniel Kammen, Director, Renewable and Appropriate Energy Laboratory.
    University of California, Berkeley, 2014.
Susy Kist, Manager of Marketing and Communications, Ocean Renewable Power
    Company, 2012.
Lisa MacKenzie, Marketing Assistant, The European Marine Energy Center,
    Orkney, Scotland, 2012.
David Plunkett, Eling Tide Mill, 2014.
Brian Polagye, Co-director, Northwest National Marine Renewable Energy
    Center, University of Washington, 2014.
Chris Sauer, Director, Ocean Renewable Power Company, 2014.
Tom Wills, Research Assistant, Aquatera, Scotland, 2014.

CHAPTER 9. HIGHER TIDES

Carson, Rachel. *The Edge of the Sea*. New York: Oxford University Press, 1962.
———. *The Sea Around Us*. New York: Oxford University Press, 1950.
City of New York. *A Stronger, More Resilient New York*. New York, 2013.
Devlin, Adam, et al. "Can Tidal Perturbations Associated with Sea Level
    Variations in the Western Pacific Ocean Be Used to Understand Future
    Effects of Tidal Evolution?" *Ocean Dynamics* 64 (2014).
Englander, John. *High Tide on Main Street*. Boca Raton, FL: The Science
    Bookshelf, 2013.
Fortis, Paolo. *Kuna Art and Shamanism*. Austin: University of Texas Press, 2012.
Georgas, Nickitas, et al. "The Impact of Tidal Phase on Hurricane Sandy's
    Flooding Around New York City and Long Island Sound." *Extreme Events* 1
    (2014).
Guidi, Ruxandra. "Will a UN Climate Change Solution Help Kuna Yala?" *National
    Geographic*, December 8, 2010, p. 17.
Hansen, James. "Scientific Reticence and Sea Level Rise." *Environmental Research
    Letters* 2 (2007), p. 13.
Howe, James. *The Kuna Gathering*. Tucson, AZ: Fenestra Books, 2002.
———. *A People Who Would Not Kneel*. Washington, DC: Smithsonian Press,
    1998.
Intergovernmental Panel on Climate Change. *Climate Change 2013: The Physical
    Science Basis*. Cambridge: Cambridge University Press, 2014.

Janin, Hunt, and Scott Mandia. *Rising Sea Levels*. Jefferson, NC: McFarland, 2012.

Kidder, Tracy. *Mountains Beyond Mountains*. New York: Random House, 2004.

Kominz, M. A. "Sea Level Variations Over Geological Time." In *Encyclopedia of Ocean Sciences*. Ed. John H. Steele, S. A. Thorpe, and Karl K. Turekian. San Diego: Academic Press, 2001.

Lane, Frederic. *Venice: A Maritime Republic*. Baltimore, MD: Johns Hopkins University Press, 1973.

National Oceanic and Atmospheric Administration. "Sea Level Variations of the U.S., 1885–2006." Technical report, National Ocean Service, Dec. 2009.

National Research Council. *Sea-Level Rise for the Coasts of California, Oregon, and Washington*. Washington, DC: National Academies Press, 2012.

Parker, Albert. "Sea Level Trends at Locations of the United States with More Than 100 Years of Recording." *National Hazards* 65 (2013).

Parris, Adam, et al. *Global Sea Level Rise Scenarios for the United States National Climate Assessment*. Silver Spring, MD: NOAA Technical Report, Climate Program Office, 2012.

Pielou, E. C. *After the Ice Age*. Chicago: University of Chicago Press, 1991.

Pilkey, Orrin, and Rob Young. *The Rising Sea*. Washington, DC: Island Press, 2009.

Pugh, David. *Changing Sea Levels*. Cambridge: Cambridge University Press, 2004.

Salvador, Mari Lyn. *The Art of Being Kuna*. Los Angeles: UCLA Fowler Museum of Cultural History, 1997.

Sherzer, Joel, ed. *Stories, Myths, Chants, and Songs of the Kuna Indians*. Austin: University of Texas Press, 2003.

Shureman, Paul. *Manual of Harmonic Analysis and Prediction of Tides*. Washington, DC: Government Printing Office, 1971.

Ventocilla, Jorge, et al. *Plants and Animals in the Life of the Kuna*. Austin: University of Texas Press, 1995.

Wood, Fergus. *Tidal Dynamics*. Dordrecht, Holland: Reidel, 1986.

Interviews

Adam Devlin, Environmental Engineer, Portland State University, 2015.

Philip Orton, Davidson Laboratory, Stevens Institute of Technology, Hoboken, NJ, 2015.

Stefan Talke, Assistant Professor, Portland State University, 2015.

GLOSSARY SOURCES

Cartwright, David. *Tides: A Scientific History*. Cambridge: Cambridge University Press, 1999.

McCully, James. *Beyond the Moon*. Singapore: World Scientific, 2006.

National Oceanic and Atmospheric Administration. *Tide Tables for Central and Western Pacific Ocean and Indian Ocean*. Silver Spring, MD: National Ocean Service, 2006.

———. *Tide Tables for East Coast of North and South America*. Riverdale, MD: National Ocean Service, 1999.

———. *Tide Tables for Europe and West Coast of Africa*. Silver Spring, MD: National Ocean Service, 2004.

———. *Tide Tables for West Coast of North and South America*. Silver Spring, MD: National Ocean Service, 2012.

Pugh, David. *Changing Sea Levels*. Cambridge: Cambridge University Press, 2004.

———. *Tides, Surges, and Mean Sea-Level*. New York: John Wiley & Sons, 1987.

Pugh, David, and Philip Woodworth. *Sea-level Science*. Cambridge: Cambridge University Press, 2014.

Index

Page numbers in *italics* refer to images.

∽ **Jonathan White** has written for the *Christian Science Monitor, Sierra, The Sun, Surfer's Journal, Orion*, and other publications. His first book, *Talking on the Water: Conversations about Nature and Creativity*, is a collection of interviews exploring our relationship with nature. White is an active marine conservationist, holds an MFA in creative nonfiction, and lives with his wife and son on a small island in Washington State.